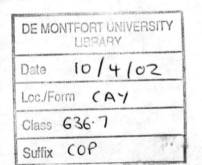

ALSO BY RAYMOND COPPINGER

Fishing Dogs

ALSO BY LORNA COPPINGER

The World of Sled Dogs

A village dog, 1974 (Jerome Liebling)

Dogs

A Startling New Understanding
of Canine Origin, Behavior, and Evolution

Raymond Coppinger
and Lorna Coppinger

SCRIBNER

New York   London   Toronto   Sydney   Singapore

SCRIBNER
1230 Avenue of the Americas
New York, NY 10020

DESIGNED BY ERICH HOBBING

Text set in Bembo

Manufactured in the United States of America

3   5   7   9   10   8   6   4

Library of Congress Cataloging-in-Publication Data
Coppinger, Raymond.
Dogs : a startling new understanding of canine origin, behavior, and evolution /
Raymond Coppinger and Lorna Coppinger.
p. cm.
Includes bibliographical references (p.).
1. Dogs—Behavior. 2. Dogs. 3. Human-animal relationships.
I. Coppinger, Lorna. II. Title.
SF433 .C66 2001
636.7—dc21
00-054137

ISBN 0-684-85530-5

We dedicate this book to eight good dogs and to Jane.

Smoky, a village dog–walking hound, selected to follow a bicycle around town all day long.

Robbie, a farmyard collie, was named after the hero of *Drei Kameraden,* because there were three of us then. He loved to ride in the back of our pickup truck, but had this disconcerting habit of growl-snapping unexpectedly at traffic cops in the middle of intersections.

Sitka, a sled dog (almost), had the best sense of humor of any dog we ever owned, and ignored all our early ambitions to be ethologists.

Scrimshaw, a small (runt, actually) Welsh corgi, had no sense of humor; but we trained her with classical conditioning techniques to "look at the wall" on command, and she thereby provided great amusement for guests at our dinner table.

Scoter, the friendly Chesapeake Bay retriever, automatically chased baseballs and brought them back to hand—but she refused to discriminate between baseballs and porcupines. She was really sweet, except when we were pulling out the quills.

Tsrna, a Yugoslavian Šarplaninac livestock-guarding dog, had only two behaviors. If you called her, she would either look at you or she wouldn't.

Lina, an Italian sheep-guarding dog, was named after Aunt Angelina Gentile, who, when we told her we'd named an Italian dog after her, burst into tears and exclaimed, "That's the nicest thing anybody ever did for me!"

Perro, a border collie. It was hard to remember that he was only a dog.

Jane, a border collie. It was hard to forget that she was only a dog.

Contents

Acknowledgments

HUNDREDS OF PEOPLE have contributed substantially to this book. Our students, who have found in dogs the same intriguing and satisfying study animal as we have. Professional colleagues, who enjoy the spirited academic debates and often offer compelling, contrasting viewpoints. The thousand-plus volunteers who tried out the novel system of protecting their livestock with guarding dogs. Sled dog drivers, retriever trainers, and a host of people in many countries who regularly ask dogs to work with them and achieve results that would be impossible without them. Physically challenged people, who rely on dogs to enhance their lives. Pet owners, who constantly confront us with the need to think harder about how dogs intersect with people.

Also, a few sled dog drivers would insist they taught us everything we know about dogs, and in the case of Charlie Belford there may be a modicum of truth to the claim. We appreciate all their input, as well as that from border collie trialers, gun-dog trainers, service-dog trainers, hound doggers, and wolf and coyote biologists, with all of whom we have had great fun working with dogs over the years. Several friends kept inviting us for winter weekends, even though we always arrived with at least sixteen noisy, demanding sled dogs. Good friends Erich Klinghammer and Günther Bloch provided forums that exposed us to diverse audiences to practice on.

Several people contributed much to the manuscript. Craig Kling and Stanley Warner read every word, trying to improve our biology-speak. Steve Weisler made us rewrite the section on differences between the various behavior disciplines. Hopeless job! Lynn Miller, although a cynophobe, is also a first-class colleague, and with characteristic energy generated critical reviews of literature that was beyond our expertise. Jerome Liebling and Kathy Doktor-Sargent donated their art, for which we are honored. Richard Schneider was the genius behind graphs we created together, but never got around to publishing.

Two people writing in the first-person plural makes the narrative awkward—as in "We fell off the sled." So, after the Preface and Introduction, we collapse into first-person singular. But all adventures and investigations were shared. As was the writing. No one knows this better than our ever-patient, ever-supportive, highly appreciated family.

The Right Kind of Dog

OUR INTRODUCTION to dogs echoed the childhood experience of millions of children in the United States. We each obtained our first dog when we were eleven years old. Within four decades, we had owned, named, and worked with on the order of three thousand dogs. Lorna's first dog was a tallish, shorthaired, floppy-eared tricolor mixed breed from a nearby farm. Ray's was a shortish, shorthaired, tulip-eared, mostly black mixed breed from a litter born behind a warehouse in the city. We acquired a jointly owned dog ten years later, when, for Lorna's college graduation, Ray presented her with a sleek, floppy-eared, orange collie-shepherd mix. During following years, several more dogs lived with us—two incredibly cute beagle pups, two incredibly uncute bloodhounds, an enthusiastic setter type, and an eager English setter. And then our children brought more dogs into our lives: a feisty little Pembroke Welsh corgi for Karyn, and a dozen Chesapeake Bay retrievers for Tim, who raised and trained them for hunting trials and field work.

The change toward a professional relationship with dogs came while we were in graduate school studying biology. Ray's official subject was animal behavior, with research into the evolution of color patterns in tropical butterflies and the reactions of hand-raised birds to novel stimuli. Lorna's was about visual acuity in birds, and how this makes them such efficient depredators of crops. But at our rural home, the pine grove was filling up with sled dogs and their houses. We had acquired a white husky-malamute female and, as happens to many incipient sled dog racers, the next thing we knew we had five sled dogs, and then ten, and then . . . at one point, including puppies, we must have had fifty dogs on hand. Over the twelve years of this competitive endeavor, we surely looked at, discussed, trained, bought, sold, accepted, rejected, and raced a thousand husky-type dogs. The conversation was hardly ever about anything but dogs. The dogs taught us a lot, but we also had a mentor in the next town, Dr. Charles "Charlie" Belford.

Belford was a veterinarian, and we'd heard of him because of his renown as a champion sled dog racer. It was due to him that we got into the sport. Because we had this pure-white, prick-eared husky that was supposedly born on a dog sled during Mush Moore's famous cross-country marathon, we figured we had a pretty good sled dog. Since Belford was a world-famous sled dog driver (at that time one of the three best dog drivers in the world), we thought he would recognize the quality and breeding of our sled dog.

We took Sitka over for her shots, and waited for the expressions of admiration—which never came. It finally had to be pointed out to Belford that this was a sled dog. How do you know? he asked, and we told him about Mush Moore, and then he asked, "Has the dog ever pulled a sled?"

If only our now dear friend could have told us then that there was no way to look at a standing-still dog and tell if it could pull a sled on the run. If he could have just told us that a running dog had to have the right running shape. If he could have told us that it didn't matter what the dog's ancestry was as long as it had the correct size and conformation. Being a Siberian or a malamute didn't necessarily mean it could run. But, no, we had to learn it all from scratch, for ourselves.

One of us decided to show Belford that the *nature* of Sitka could be *nurtured,* and trained her to pull a sled. That is when the yard began to fill up with dogs because it turned out that Sitka needed help, and the various new purported sled dogs (they were purebred Siberian huskies) we acquired were about as good as she was and it took a lot of them to pull a sled—and not very fast. Within a few years, Ray got his Ph.D. and his first real job as a professor at Hampshire College. At the same time he began switching from Siberians to the Alaskan huskies that hadn't made Belford's team. The sixteen-dog team he slowly developed over the years eventually ran as fast as any on the New England/Canada circuits. Sometimes they were even number one. Karyn, too, raced a five-dog team, and she helped add to the growing number of trophies. Ray's team eventually achieved a distinction in racing circuits during the 1970s, because of its leader, an unusual border collie named Perro. Perro was a sort of gift from a statistics professor at nearby Amherst College. He and his family were reluctantly giving away their pet dog because he chased cars all day. Perro turned out to be a dog driver's once-in-a-lifetime lead dog. He was taller and lankier than most border collies, very fast, and he kept the Alaska husky team dogs on the right trail, responding immediately and accurately to "gee" or "haw." Sitka had long before been outrun, and gracefully retired.

Part of the fun with the sled dogs was discovering what makes a good sled dog. Why were purpose-bred sled dogs like Belford's Alaskans better for the job than the other dogs people sometimes hitched up—the pointers, the Irish setters, even the Siberians, for instance? Was it the nature (genes) of the dogs or the nurture (developmental environment)? Our hobbies and our professions blended as we began to ask questions about sled dogs. For example, why did some dogs, but not all, form debilitating little "snow balls" of ice crystals between their toes during a race? We designed an experiment comparing footpad sweating of Siberian huskies, coyotes, and wolves, and found out that the domesticated canines and western coyotes had a greater density of sweat glands in their pads than wolves and eastern coyotes. We also asked, How does ambient temperature or training affect a dog's ability to run? We measured temperatures in running sled dogs, and determined that at ambient temperatures over 60 degrees Fahrenheit, running sled dogs cannot radiate enough heat to counteract the internal heat buildup. Above 60 degrees, they overheat and become ineffective runners. These results provided information about microanatomy and physiology of sled dogs and demonstrated clearly how well adapted they are to their specific environment. We published the results in scientific journals. Ray adjusted the management of his racing team, to the benefit of the dogs. Meanwhile, Lorna finished her degree and was studying the history of sled dogs and taking photographs at the races. She wrote the first comprehensive book about the use of sled dogs for moving freight, aiding exploration, and racing for sport, and about the men and women who drove dogs and achieved success as dog handlers.

Then, during a sabbatical leave from Hampshire College, we began a dog behaviorist's dream project. The sheep industry needed a new idea for protecting sheep from predators, and the researchers at the Winrock International Livestock Research and Training Center in Arkansas were trying to find out if it was really true that in Europe, special breeds of sheepdogs lived among the flocks and protected sheep and goats from wolves, bears, and human thieves. In the United States at that time, "sheepdog" meant a herding dog. Winrock liked the work we had done in pinpointing aspects of breed differences in sled dogs and asked us to try to sort out the sheepdogs. They sent us on a quest, which began with visits to farms and ranches all across the country. Karyn, newly licensed to drive, and Ray took the sled dog truck and drew a convoluted track across the map of the United States. We covered ten thousand miles during March 1977, and searched out anybody who claimed to know anything about sheep-guarding dogs. There were only about twelve serious stops.

In Texas, we spent days following a komondor, which is a Hungarian flock-guarding dog. Maggie lived with and was in charge of hundreds of goats, and she convinced Ray and Karyn that she was truly protecting those Angoras from the ever-present coyotes and bobcats. We told Winrock that if you could get one dog to do it then you could get a thousand—if you knew what you were doing, and if you had the right kind of dog. On the trip we managed to talk a cowboy out of Bandit, a cross between a border collie and a Queensland blue heeler cattle-herding dog. The cowboy had taught Bandit that the softer he talked, the madder he was, and we were able to impress our friends with how good the dog's hearing was by whispering commands over fairly long distances and having her respond immediately.

In those days there was a healthy skepticism about having big dogs in the same pasture with sheep—understandably so. One Texas rancher had enthusiastically bought a komondor or two, but had run into trouble with them chasing and biting his sheep. He told us it was bad enough when he got his predators for free, but he was darned if he was going to pay for them. Our experience, too, from childhood books such as *Bob, Son of Battle,* told us that dogs chased sheep and ate them. Our sled dogs never killed anything but the best chicken in the neighbor's flock.

But we had seen Maggie, and so, with the help of Winrock and Hampshire College, we journeyed to the sheep pastures of Yugoslavia, Hungary, Switzerland, Austria, Italy, France, Portugal, and Spain, and even, with additional help later on from the Anatolian Shepherd Dog Club of America, to Turkey. These were supposed to be fact-finding trips, but like always, we weren't able to just look. We bought four pups in Yugoslavia and shipped them to Lorna's mother in Boston. She turned them over to Ray's cousin Barry, and he added them to the four we had earlier shipped him from Italy and the two from Turkey. When we returned from Europe that first summer (via Scotland, where we had picked up six border collies), we found the four Italian Maremmano-Abruzzese pups, the four Yugoslavian Šarplaninacs, and the two Turkish shepherd dogs doing just fine. We kept all sixteen puppies in the same pen, and the question was, What is the behavioral difference between a herding and a guarding dog, and when during a pup's life do these differences appear?

We knew, of course, that herding dogs were bred to herd sheep and guarding dogs to guard sheep, but we were after the deeper differences between them. What were the differences in their brains that separated so markedly the herding from the guarding behaviors? *Why* were they different?

We had sold the sled dogs. Twelve years of fall training, winter racing, and spring and summer maintenance were enough. It was time to move on to some new, exciting research.

Between 1977 and 1990 we bred and placed more than fourteen hundred livestock-guarding dogs on sheep farms and ranches across the United States and Canada. The original stock was augmented during successive trips to Europe and Turkey. Some of our original grants were extended, and the U.S. Department of Agriculture's Science and Education Administration added three years of support. We kept track of every dog, recording their successes and failures for fourteen years. We were on the road constantly, observing our dogs, camping in high mountain pastures with the flocks, giving talks in grange halls. As part of our academic research, we began studying behavioral variation within the guarding dogs, pondering why some dogs were attentive to sheep and others were not. We also wondered what mechanisms in dogs led to the creation of two very distinct kinds of sheepdogs: those that herd and those that guard. They both worked with sheep (and often, other kinds of livestock), but they looked and behaved very, very differently. Was it training, developmental environment, or genetic? Was it nurture or nature?

Sheepdogs are an exceptional subject for behavioral ecologists, or ethologists. Here are two distinct races—herding dogs and guarding dogs—within a single species (*Canis familiaris*), both chosen to work in the same environment (grasslands), both chosen to respond to the same environmental stimulus (sheep), but their genetic natures call for them to respond to sheep in two distinctly different ways. A biologist has to wonder how two races within the same species have evolved to behave so differently. Aside from variations in coat color and size, these two types behave as if they are "wired" in highly differentiated ways.

There is probably no other place in nature where such a perfect situation for comparative study within a species exists. Researchers commonly have trouble sorting out what parts of a behavior belong to the genetic nature of the animal and what parts are a product of development and nurturing. Designing experiments to compare two breeds is difficult, because one of the breeds is always in the wrong environment. If we were trying to find out if the difference in behavior is genetic, we have to observe the two breeds in the same environment in order to eliminate environment as an experimental variable. But if they are adapted to different environments, then one of them will always show atypical behaviors because it is out of its natural environment.

In studying the sheepdogs, we discovered why dogs are to us one of the most fascinating animals on earth. Here are breeds within a single species, animals that have no appreciable differences in their genetic codes, with no infertility between the breeds, and yet they innately learn different things as they mature, even in the same environment. The six border collies and ten livestock-guarding dogs we bought in 1977 were all born in their native countries on or about April 19, transported to our yard before they were eight weeks old, placed in the same large paddock, and cared for by assorted adults, children, and students. No sheep were present. By the time they were ten months old, the border collies were acting like a different species, showing their breed-characteristic "eye-stalk" behavior and trying to herd each other. The guarding dogs played with each other and ignored the collies. None of the guardians ever showed eye-stalk. It was a powerful example of the effects of small genetic differences.

By 1990, we had stacks of data about the two types of sheepdog. We had observed sheepdogs in the farm kennel-laboratory at Hampshire College, out on sheep pastures across the United States, and again in the Old World in their original habitats. We published papers and reports for the scientific and the farming communities. After winding up the applied aspects of the research, we began to look more closely at theoretical and practical implications of what we'd found.

What stood out, in particular, was the fundamental contrast between the two types of sheepdogs. Each showed distinctive behaviors, and the more we studied them, the more we learned about the relative effects of genetics and environment on their adult behavior. We began to see the absolutely critical relationship between a pup's early development and its abilities to learn or perform a task in adulthood. The reasons for subtle differences in breed behaviors began to become obvious. We think that in this book we can explain these differences and the reasons for them, and thus add an essential dimension to information about dogs.

Many books have been written about dogs: choosing, raising, training; their intelligence, their health problems, their psychiatric problems, their inner lives, their secret lives, and their love lives. They delve into the dog's domain and try to explain their behaviors. There are good books on dog training, and we know some super dog trainers. Both provide techniques for training, or for modifying a problem dog's behavior, so that the dog and the people in its environment will enjoy each other. But dog training is commonly done using a technique called the "conditioned response." It is based on punishment/reward. The animal is punished for a wrong response, rewarded for a correct one. It is basically the same method used to train whales, or rats, or pigeons. The vocabulary and the approach are rooted in the field of psychology—the study of the mind and mental processes.

But dogs are not whales or rats, and especially they are not pigeons; they do not respond to instrumental conditioning in the same way. They are a beautifully different organism. But what is that difference? How are they different? Why? How did they get different, and does it really matter?

Some people who work with dogs know the answers to some of these questions. Bird dog trainers know they can't train a young dog until after it shows "point." Point is innate; it is wired into the dog's brain. The same is true with the border collie, which has "eye" hard-wired in. It's impossible even to try to herd sheep with a pup until it shows eye. What a border collie handler does is train the dog when and how to go in order to use the eye. But nobody can train a dog to show eye, or to point.

A few years ago, people who work their sheep with border collies waged a campaign against the registration of their breed by the American Kennel Club. Because showing eye is a genetic trait, it can be selected against, and if border collies are going to be bred for shows and as pets, breeders are going to have to get rid of the eye. We got Perro because he showed eye, stalk, and chase behaviors all day long to passing cars, chasing them up and down the suburban street. A dog that shows eye can be a problem pet. On the other hand, breeding border collies for pets tends to winnow out those with intense working behavior. One can't have it both ways, as we shall see. That is why working-dog people prefer to keep their top-flight working dogs out of the pet and show markets.

Those trainers and handlers are showing a basic understanding of a dog's behavior that is rooted in biology and cognitive psychology. They understand that dogs are biological organisms, growing and acting within biological constraints.

Other people who work with dogs and use standard instrumental techniques often come to realize that something is missing. Trainers of dogs to assist physically challenged people, for example, are well aware that over half the dogs that enter training do not finish, or do not qualify at the end of training for their new jobs. They think what they need is a "better" dog. Pet owners complain of dogs that have behavioral problems: they chew furniture, bark all night, have "anxiety attacks." Could it be that many of these problems could be solved by paying attention to early development of pups, by exposing them early and consistently to the environment and events of their adult life?

Very few dog books have been written by biologists. What we have learned since we took Sitka to Dr. Belford's office has made it clear that dogs need a new book, based on their biology, explaining what they are, why they are different from wolves and each other, how they got that way, and how their relationships with people can be enriched so both species benefit. We hope this book will fulfill that need.

We have tried to treat all the dogs in this book consistently. But we realize that even though we are supposed to be unbiased, objective scientists, we are also two of the world's most avid admirers of dogs. Our prejudices may show through and we may make mistakes. If our preconceptions are unfounded, we apologize. If the controversial portions generate discussions that lead to improved relationships with dogs, then we will suffer all criticism, and admire the dogs (and their human symbionts) even more.

Studying Dogs

WHY STUDY DOGS?

As biologists, we would characterize the species domestic dog—*Canis familiaris*—as successful.[1] Biologically speaking, they are incredibly successful. What this means is they have made the transformation from their ancestral form, the wolf, and exist today as domestic dogs in hugely significant numbers. Dogs as a species are most likely less than fifteen thousand years old, which is a barest instant of evolutionary time.[2] Wolves as a species are maybe five million years old, and they need protection from extinction.

Domestic dogs are common and ubiquitous over most of the world. Their wild relatives in the dog (Canidae) family also live abundantly, but in much more limited areas and in much fewer numbers. A ballpark estimate of the relative numbers in the world shows this discrepancy. (See the table on page 22.)

Four hundred million dogs in the world—that is a thousand times more dogs than there are wolves. If wolves are the ancient ancestors of dogs, that means dogs have achieved a biological coup, successfully outpopulating their ancestors by *a lot.*

For the last hundred and fifty years, discussions about which of the wild canids the dog descended from have attracted many notable investigators. We will tackle that question briefly at the end of this book. But a "correct" answer really doesn't matter for this story. We use the term

[1] For a discussion about the recent reclassification of the dog as a subspecies of the wolf, *Canis lupus familiaris,* see Chapter 9.

[2] For a discussion about the recent news that dogs might be 135,000 years old, see Chapter 10.

Populations of Wolves, Coyotes, Jackals, and Dogs

Wolves *(Canis lupus, C. rufus)*	**400,000**
Coyotes *(C. latrans)*	**4,000,000**
Jackals *(C. mesomelas, C. aureus, C. simensis, C. adustus)*	**40,000,000**
Dogs *(C. familiaris)*	**400,000,000**

Relative populations, worldwide, of the genus Canis.

"wolf" for the ancestor of the dog because it is short and easily recognized. If the original dogs didn't descend from the animal called *Canis lupus,* they descended from something very much like it.

When one species evolves into another, often the evolving species is better adapted to the changing niche than is the old. The new species thrives and the old species passes into fossil land. Areas previously occupied by wolves are now occupied by dogs, living in changed habitats. Dogs now have a greater distribution worldwide than the wolf ever did.

Biological success can be quantified. We can measure the success of dogs by counting them, or by weighing their populations. Biological "success" simply means something like: there are many more pounds of dog in the world than there are pounds of wolf. In North America, for example, there may be fifty thousand wolves, all but 5 percent of which live in Canada. But there are on the order of one hundred million dogs in North America (United States, Mexico, and Canada). That is *two thousand times* as many dogs as wolves.

In the United States, our colleagues estimate fifty-two million pet dogs. That implies one hundred and thirty times as many pet dogs in America alone as there are wolves in the whole world. We count wolves, worldwide, in the hundreds of thousands. We count dogs in the hundreds of millions.

Nobody knows the exact figures of any of these species. We may have underestimated some and overestimated others. But we bet that when someone counts them exactly, by means of satellite imagery, perhaps, there will be a staggering number of dogs compared with wild canids. Dogs could well be the most populous carnivore with the widest geographic presence.

That is an evolutionary success story. But which evolutionary story? Is it a Darwinian process?

The fact that dogs sprang from wolves as a distinct form probably no more than fifteen thousand years ago is a miracle of evolution, requiring an extraordinary evolutionary process. Dogs—like some fantastic chameleon—have taken on an endless variety of shapes. Dogs may well display the greatest range of shapes of any mammal that has ever existed. As reproductive adults, they may have a greater range of sizes and shapes than any vertebrate species that ever lived. Why is that?

Darwin thought that evolution proceeded from the simple to the complex. Each new species was more sophisticated, more specialized than the one it replaced. It is certainly true with the canids. The wolf has a basic predator shape. It is not a specialist, but rather a generalist. In comparison with other large carnivorous predators, wolves can't run very fast; they have relatively weak jaws; they don't see very well. The cat family runs faster, bites harder, and sees better than the dog family. Some would argue that wolves have the penultimate social behavior, exceeded only by our own. But that is an overstatement; many carnivores, ungulates, and even rodents have social behaviors just as, if not more, complex.

Wolves and their relatives haven't changed very much if at all in the last five million years. In fact, some experts have averred that the canids haven't improved much in four million years. They are an effective, widespread, and basic predator. They are not very good at adapting to change. Wolves are often referred to as an indicator species, which means that any little deterioration of their habitat causes an immediate drop in their numbers in that habitat. They don't seem to be able to adjust to expanding civilization the way coyotes do. The coyote's range is increasing in the face of human expansion, while the wolf's is decreasing.

Dogs have shapes and behaviors unheard of in the wolf. Unlike the wolf—if we have seen one wolf, we basically have seen them all—dogs are continually interesting. Take any specific behavior and there is a breed of dog that can outperform any wolf. Compared with wolves, sled dogs can run farther, greyhounds can run faster, bloodhounds have a better sense of smell, borzois have more optical overlap and better depth perception. Some would contend that, cognitively speaking, the wolf is smarter. That may be, but we would propose that if it is so smart, why can't we teach one to herd sheep, or fetch a ball, or deliver a bird to hand, or guide a blind person through the crowded streets of a city?

The message we're going to emphasize in this book is that the dog is first and foremost a biological being, and no mere subspecies of the wolf. Each individual dog represents an end product of a whole series of significant biological events. Our dog lying out there under the car on a hot day is not an evolved wolf in its simulated cave. It is a highly evolved, specialized new animal with behaviors that adapt it to its niche.

But when books about dogs begin, "Dogs are closely related to wolves . . ." and then go on to equate many dog behaviors with wolf behaviors, we wonder how such a superficial assessment would work if a book about people started out, "Humans are closely related to chimps. . . ." Dogs may well be closely related to wolves but that does not mean they behave like wolves. People are closely related to chimps but that doesn't make us a subspecies of chimpanzees, nor does it mean we behave like chimps.

Another problem in understanding domestic dogs is the tendency of researchers to see domestic animals differently from wild forms. They look at domestic species as if they had gone through some invalid kind of evolutionary process. Perhaps they feel that the dog's adaptation and evolution were not "natural," and so are inferior. Many scientists also discuss domestic animals as though they are degenerate forms, not worthy of serious study. We suppose they may think this because they see domestic animals as having lost the "normal" biological traits that enable them to survive in the wild. This distinction is underlined in our educational institutions. At the university level, animal science deals with domestic animals, and is usually a department within agriculture, not zoology. Animal science traditionally has not been very interested in behavior; and zoology doesn't think chickens are real birds. Animal science rarely even teaches about dogs. Dogs are almost exclusively the province of veterinary medicine, and even there little attention is given to their behavior. Those of us fascinated with dog behavior are caught between disciplines. We are anomalies in all academic fields.

Where does information about dog behavior come from, then? It is pieced together by people in a variety of fields, such as animal science or animal psychology or grass-roots dog training. Rarely do we find academically trained biologists focusing on dog behavior. And yet dogs carry important clues about mammalian behavior, especially about its inheritance and adaptation. What is truly intriguing about dogs is how quickly the selection for performance of a particular behavior can lead, usually inadvertently, to a distinctive new form. That form can carry out that behavior better than any other dog (or breed or species). No other dog has ever been demonstrated to herd sheep as well as a herding dog. No other terrestrial animal can run marathon distances as fast as a sled dog. A team of sled dogs running efficiently across a snowy landscape is one of the most beautiful sights in nature. These are not degenerate forms. The refinement of a dog's behavior is exquisite. Any breed of dog behaves with much greater complexity than any wolf. Therefore, in Darwinian terms, dogs should be viewed as an evolutionary advancement, worthy of study in all the traditional biological disciplines.

Although we realize that dogs are phenomenal, most people don't seem to realize how phenomenal. People rarely have enough information to put dogs into an accurate perspective. While racing sled dogs, we would constantly be asked if our dogs had wolf blood. We have been told many times that Eskimos breed their dogs with wolves to make them better, stronger, or faster. Such comments signal to us a lack of appreciation for what a dog really is. The speaker is treating dogs as if they were an inferior form (of wolf, we guess) that could be improved by breeding with wolves. Such a speaker cannot be cognizant of the fact that it is sled dogs that are superior in this context. To breed a sled dog to a wolf would be at least an evolutionary digression, if not a degradation in the behaviors of both. Wolves are not sled dogs; they have not been selected to be sled dogs. They cannot perform that task as well as sled dogs. Why on earth would we want to breed our super sled dog to something that not only lacked the specialized form for racing, but would introduce behavior problems to the team as well?

As we look at adaptations by dogs, we also begin to recognize that the relationships of dogs to humans is not all that clear. The old clichés and platitudes such as "man's best friend" don't really hold for dogs all the time. When we explore the social relationships between dogs and people in biological detail, we get a startlingly different picture from what is usually portrayed. "Man's best friend" is not an ecological definition; that is, it does not define the relationship between dogs and their environment. What about the inverse: Are humans dog's best friend? To be true behavioral ecologists, we should define the terms and test the equations. The name of the biological game is survival. To survive, an animal must eat, avoid hazards, and reproduce. How does delivering the bird to hand benefit the dog? What does working hard pulling a sled get for the dog? Is it to a guide dog's biological benefit to be chosen for this ostensibly humanitarian role? Are dogs simply slaves that have been trained for robotic tasks? Or is there an inherent reward here for dogs?

We think the answer to some of these questions might be negative, and in many cases it might not be to a dog's biological benefit to enter into a symbiotic relationship with people. Why would people cringe, thinking about opera-singing castrati giving up their reproductive potential at a young age so that they can maintain a beautiful singing voice for the pleasure of the listeners, and not also cringe at mutilating a dog in the same way for the purpose of maintaining its docile juvenile behavior for the benefit of a blind person?

As we travel the world studying dogs, we see the variety of relationships dogs have with people. Several very different relationships are obvious. The majority of the world's dogs still live as wild animals, as scavengers of villages, with people paying little or no attention to them. Some pastoral dogs actually do have a mutual relationship with their shepherds. Other dogs parasitize humans, taking their food and endangering them with bites and diseases. In still other cases, in the name of "breeding pure" and "improving the breed," humans have captured and caged entire populations of dogs and bred them into bizarre shapes, often creating freaks. These latter dogs lead impoverished lives. They can run or play only at their human's pleasure, and some are physically incapable of doing so. Their very shapes predispose them to lead painful and sickly lives. Many a dog is born in a kennel, in a cell, mass-produced for a commercial venture. Some of these are raised to test the virulence of pathogens or cancer cells. For these best friends, only their tissues work for humankind.

It is obvious to us that of the four hundred million dogs in the world, only a tiny percentage have a truly mutual relationship with people. Only a few cooperate in some task, as suggested by the "boy and his dog" image. We are not proposing that such relationships with dogs are necessarily wrong. We just think we should not deceive ourselves about our actual alliance with dogs. To continue current breeding programs that close a breed's gene pool to any new genes is in many ways cruel to the dogs. Many pure breeds already suffer physical deformities, some selected for on purpose, some resulting from closed gene pools. Mental deformities are also evident, particularly when people ignore or neglect the early development of their dog.

In this book, we'll explore the nature of the various symbiotic alliances, so the reader can gain a biologically informed perception of what people are doing to dogs. We intend that this view of dogs could help to cultivate a better relationship between our two species.

HOW TO STUDY, AND WHO STUDIES, DOGS

Populations of animals can be studied in different ways by different "-ists." When ecologists speak about how people and dogs relate, they use a different vocabulary than when economists, say, explore economic questions of how they relate. People tend to assume that the relationship between people and dogs is good, of mutual benefit to both species. People supposedly gain value from owning a dog, or perhaps the dog does valuable work for them, and they reward it by vaccinating it against disease, protecting it from harm, and providing it food and reproductive opportunities.

But, ecologically, the idea of mutual benefit ceases to be an assumed truth and raises the questions: Is the symbiotic relationship between humans and dogs really mutualism? What else could it be? The ecologist defines four basic symbiotic relationships. Mutualism is only one quarter of the story.

Commensalism is a symbiotic relationship that is good for one species but does nothing for the other. There are millions of dogs around the world scavenging the dumps of villages. They get a food benefit from living close to people, while the people get little or no benefit from the dogs. Some people would contend that the scavenger benefits the village by cleaning up the refuse. In that case, rats and raccoons should be given similar credit. There are so many dogs making a commensal living around civilized people that we think villages may well have been the original inspiration for wolves to become dogs.

Mutualism is the relationship assumed to exist now between dogs and people. The first thing people point to, to support this, is that dogs pull sleds, herd sheep, or guard houses. With our varied experiences with working dogs, we wouldn't deny this. But others would say that it is cruel to have a dog pull a sled, implying that it is not to the dog's benefit. Certainly, and by definition, if there is a mutual benefit it has to benefit both parties, not just one. A major portion of this book is devoted to dogs that work with people, and we will demonstrate why some of this work is mutual and some is not.

Parasitism defines a relationship between two species living together where one organism obtains a benefit at the expense of the other. It may be unpopular, but we are going to make the case that the domestic house dog may have evolved into a parasite. It costs more than it gives back. Further, we postulate another relationship, a subcategory of parasitism, called *dulosis*. Dulosis is enslavement, where one species captures workers from another species. We are resigned to the fact that we will probably lose this argument, but we bet we lose it because people just don't agree with us, and not because someone can provide data to the contrary.

Amensalism. There is another way we might lose the parasite argument. The fourth form of symbiosis is *amensalism,* a living together in which one species hurts another, often unknowingly and without benefit to itself. If we can show that pet dogs are in a peculiar position where they are genetically trapped in small, inbreeding populations (called pure breeds) that will eventually destroy them, it should be clear that it is bad for the dog to be in this relationship, and not particularly beneficial for the human.

The presence of parasitism or amensalism should not be taken as grounds for the elimination of dogs in our lives. Rather, the facts should be used as starting points for change and moving toward real mutualism.

Throughout the world are populations of dogs living together with human beings in one or another of these symbiotic relationships. But there is an additional wrinkle for the dog. The symbiosis (whichever one it is) with people is *obligatory* for the dog. "Obligatory" here is a biological term. An obligatory parasite is an animal that has to be a parasite on one species of animal. For example, the roundworms parasitizing a dog cannot live in a human, or a cat, or any other species. Their specific adaptation obligates them to live in the canine intestinal environment.

Some symbiotic relationships are obligatory for both species. The nitrogen-fixing bacteria on alder trees is an example. Both species need each other in order to survive. But this is not true in the dog-human symbiosis. Domestic dogs cannot survive outside the human domain. But humans can survive without dogs.

Only a very few of the six billion people in the world depend on the dog for anything. If dogs disappeared from the face of the earth tomorrow, humans would survive the tragedy without much stress. But if humans disappeared tomorrow, dogs would likely become extinct shortly thereafter. We suppose if humans disappeared slowly, domestic dogs might evolve into something like the dingo. But evolve they must because it is doubtful they could compete in a wild world in their present forms.

The fact that dogs are obligated to live with people means that people can exercise a certain power over them, and can force them into any kind of relationship. Many of these relationships are as heartwarming as anyone could possibly want. Others are difficult for both the people and the dogs.

A variety of specialists are interested in dogs. They are ethologists, behavioral ecologists, cognitive psychologists, neuroscientists, biological anthropologists, developmental biologists, and behaviorists. At universities, these researchers are usually associated with departments of zoology, psychology, or anthropology, as well as animal or veterinary science. And recently, a new discipline has evolved, anthrozoology.

Actually, given the 400,000,000 dogs in the world, very little research on dog behavior takes place at the university level. Much of what we deduce about dogs is borrowed from studies of other species. What a neuroscientist discovers about octopus nerves applies to human or dog nerves. The eyes of one species of mammal work the same way as another. If you know how wolves see, then you have a pretty good idea how dogs see. This is the principle of homology, which assumes that organs with an evolutionary or phylogenetic relationship will behave similarly.

Zoologists, psychologists, and anthropologists all believe that behavior is adaptive, in an evolutionary and in a developmental sense. Behavior is a synergism between the genetic nature of the animal and the environment in which it finds itself. This synergism is nature *times* nurture, and not, as was argued until the 1950s, nature *plus* nurture. Donald Hebb said it clearly in 1953, that trying to figure out which was more important, nature or nurture, was like trying to determine whether length or width was more important in understanding the area of a field.

Even though animal-behavior students in different departments tend to agree on the fundamental tenets, they also tend to focus on different parts of the equation. Animal psychologists are interested in behavioral development. They are to psychology what embryologists are to zoology, in that the psychologists study the development of behavior while the embryologists look at the development of physical structure. Biologists often focus on how the animal behaves in a particular environment. Anthropologists tackle the history and evolution of the symbiosis of animals with people. In the dog world there is tremendous emphasis on applying the findings of the laboratory to real-world problems. Thus, veterinary schools are more likely to employ clinical psychologists than ethologists, and a vet who applies behavior principles to specific problems might identify himself as an animal psychiatrist. Animal behavior is complicated, and no single approach is adequate to the task of unraveling a behavior.

When animal behavior emerged as a discipline in the middle of the twentieth century, some bitter battles were fought over the various approaches and the meaning of the results. These were centered on the nature versus nurture controversy. In the 1950s and 1960s, American psychologists and European ethologists argued heatedly. When the European ethologists Konrad Lorenz, Niko Tinbergen, and Karl von Frisch were awarded the Nobel Prize in 1973 for (essentially) inventing the field of animal behavior, there was resentment on this side of the Atlantic among the pioneers in the science of behavior. And well there should have been!

At present, theoretically, there are no fundamental differences between anthropologists, psychologists, and biologists. Why resurrect dead issues? A new understanding could help solve significant problems that persist in the dog world. Between academic behaviorists and those trainers of dogs in the workaday world, a gap exists. Although the popular press tries to bridge the gap, it is very hard to do, and old myths keep creeping back into the prose. Perhaps the paucity of studies specific to dog behavior has hampered our understanding of this species. Perhaps too much of what we infer about dog behavior is really carelessly borrowed from studies of other species such as wolves or rats. Perhaps reporters have been too ready to accept an exciting new announcement from the laboratory, not realizing that even scientists can misinterpret data or draw shaky conclusions. We think it is amazing when geneticists discuss breeds of dogs as if they were ancient, sexually isolated species, artificially selected in the Darwinian sense for the service of people. People who should know better write about genes for behaviors such as water-loving in Newfoundlands, or herding in border collies. Professional dog behaviorists will uncritically homologize: wolves form packs, and therefore if you want to teach a dog a trick you have to be the pack leader, the alpha wolf.

Part of the problem is nestled in the history of animal behavior itself. Perhaps a 1950s clarification of what ethologists and psychologists were thinking about dogs might help.

Ethologists asked, What *do* animals do? They were interested in species-typical behavior, often called innate or instinctive. They assumed that a species' behavior was an evolutionary (genetic) adaptation. Ethologists constructed ethograms, an inventory of an animal's motor patterns. Motor patterns are a posture assumed by an animal. John Fentress and Peter McCloud, at Dalhousie University, said it best in a 1986 paper: "It is through the production of integrated sequences of movement that animals express rules by which they interact with, and adapt to, their physical, biological and social environment."

Two fear motor patterns displayed by dogs to an approaching stranger illustrate integrated sequences of movement expressing dog rules. A male dog raising his hind leg to urinate exhibits motor patterns that appear in the dog's ethogram, but not in the cat's. Similarly, it appears that livestock-guarding dogs and herding dogs have different sets of rules, which is why they behave differently in the presence of sheep.

Some would argue that the original ethologists evolved into the modern behavioral ecologists. Others of us believe (and hope!) that not all ethologists have gone extinct.

Behavioral ecologists study how a species of animal earns a living. How does the animal behave in order to capture energy (food), and how does it translate that energy into offspring? The behavioral ecologist is interested in defining a species' niche. One species, one niche, so to speak.

Fear motor patterns. Showing a fear motor pattern, this dog intends to flee from an approaching stranger. The posture a dog assumes gives us insights into what its intentions are.

This dog is also showing a fear motor pattern. In this case, however, the dog is undecided whether to fight or flee from the approaching stranger. (Drawings by Katherine A. Doktor-Sargent from photos by the authors)

A behavioral ecologist studying dogs assumes that its species-typical behaviors are adaptive. An adaptive foraging strategy for dogs is to display care-soliciting motor patterns, causing humans to feed them. If humans don't profit from feeding dogs, then the behavioral ecologist classifies the dog as parasitic on humans. It doesn't matter that the human likes to be parasitized by dogs. The question of whether people feel good feeding dogs, or get some psychological benefit from feeding dogs, is not within the methodology of the behavioral ecologist. For that we must turn to the psychologist.

Animal psychologists are more interested in the development of behavior, whether it is species-typical or not. During the 1950s, there were some misconceptions about what animal psychologists actually did. There was some sense that psychologists didn't believe in species-typical behavior, and discounted behavioral genetics, emphasizing that development was nurtured. It was never true. They don't deny species-typical behavior, nor do they deny a genetic component to behavior; they are just interested in the process of development. Another misconception was that they were interested in the modification of behavior. For one subset of psychologists, the behaviorists, this was almost certainly true.

The behaviorists did not ask the species-oriented question, What *does* an animal do? but rather, What *can* an animal do? The domain of the behaviorist was nurture, or learning. Animals learned to adapt to their environments. Behaviorists watched rats in mazes, or trained pigeons to play Ping-Pong, or studied the abilities of chimps to use tools. It was easy to think of them as modifiers of behavior because of all the popular discussions about Pavlov, Skinner, and Freud.

Dog behaviorists, that is, dog trainers, fall into two overlapping fields. They are Pavlovians or Skinnerians. Pavlov rang a bell and then fed his dogs. Later he noted that when he rang the bell, the dogs began to salivate in anticipation of getting fed. "Click and treat" is the modified training technique based on this so-called classical conditioning. The dog associates the click with the presentation of a treat and responds appropriately even if no treat is present.

The Skinnerians modify behavior by rewarding proper behavior and/or punishing improper behavior. This technique is called operant and/or instrumental conditioning. If the guide dog walks the trainer into a pole, then the trainer scolds the dog, giving a quick tug on a choke collar (the instrument). What the dog learns is to avoid those situations where it gets punished (aversive conditioning).

Animal psychologists or animal psychiatrists often work on aberrant behavior, which they describe in psychological rather than ethological terms. This dog has separation anxiety; that dog has a compulsive disorder. Since many of these behavioral disorders don't respond to classical or operant conditioning, then the specialist might prescribe drugs.

Our complaint here is that behaviorists tend to think all animals learn the same way. What works for a dolphin works for a rat works for a dog. What works for one breed of dog works for another. If one breed of dog trains better or easier than another, it is easy for the behaviorist to jump to the conclusion that the one breed is more intelligent than the other. But such easy leaps do not take into account that dolphins and rats and dogs and the various breeds thereof are all working with different sets of rules. And this is where the gap in understanding dog behavior lies.

The following story may be apocryphal, but it contains the essence of the problems facing dogs and their owners. A pet owner calls the veterinarian/behaviorist about his dog because the dog is acting strangely. It stands or lies absolutely still for long periods, staring at an object, and can hardly be distracted. The vet observes the behavior, and diagnoses an obsessive-compulsive disorder. His remedy is medication. But what kind of dog is it? A person experienced with border collie behavior understands that the dog is showing "eye," a perfectly normal and highly desirable behavior for this breed. This motor pattern is genetically hardwired, and without it the border collie cannot be trained to control livestock. The behavior cannot be cured with a drug. Nor can it be cured with training. It is part of the dog's ethogram. To attempt to treat the animal for a psychological disorder stresses both dog and owner. To keep such an animal as a pet also stresses both dog and owner.

Groundbreaking research on dog behavior in the United States was done during the 1950s and 1960s at the Jackson Laboratory in Bar Harbor, Maine, under the direction of John Paul Scott and John Fuller, behavioral psychologists. Scott and Fuller were motivated by the question, What effect does heredity have on behavior? Both zoologists and psychologists worked in their laboratory, trying to integrate psychology and behavior genetics to answer the question. Publications from that extended project often used the vocabulary of psychology. This made sense, because although dogs were their experimental subjects, it was human social behavior that motivated the research. Scott and Fuller's 1965 report on this Bar Harbor project was published as a book, *Genetics and the Social Behavior of the Dog*. It has been reprinted and reissued several times, because people interested in dog behavior are—as they should be—fascinated by it.

Konrad Lorenz's book *Man Meets Dog* was perhaps the first ethological approach to dog behavior. It is a delightful book for a dog lover, published in 1954. In it, his analysis of breed-typical behavior led him to hypothesize that some breeds descended from jackals, while the northern breeds like the spitz types descended from wolves. The "lupus" dogs, he claimed, treat their masters as pack leaders, and are independent of personality. The "jackal" dogs treat their masters more like parent figures, and show "slave-like submission."

We were lucky enough to meet Lorenz in 1977, and he greeted us from the steps of his home in Altenburg, Austria, with the words, "So you are the dog biologists. Before we start our talks I just want to say, everything I've written about dogs is wrong." He was referring to his ideas about the wolf and jackal dogs. Well, one rarely hears such an admission, but as a scientist, Lorenz was a great hypothesizer, and as a truly great scientist, he was not afraid when colleagues could refute his ideas. We like to think that ethology really started in 1937, when Lorenz wrote, "behavior is a taxonomic character." In other words, behavior is a diagnostic trait. For ethologists, dog behavior is a diagnostic characteristic of dogs.

If *Man Meets Dog* was the first popular biological approach to dogs, and *Genetics and the Social Behavior of the Dog* was the second, then Michael Fox's *Canine Behavior* (1963) and *Understanding Your Dog* (1972) come next. Fox, an ethologist, veterinarian, and psychologist, worked with Scott and Fuller at Bar Harbor, and his books are solidly grounded in the biology behind the behavior. Fox was the first to publish popular books that integrated comparative psychology and ethology. He was the first to design experiments listing motor pattern differences between domestic and wild canids. He was also the teacher and mentor of a later generation of dog investigators, notably Marc Bekoff at the University of Colorado and Randall Lockwood at the Humane Society of the United States.

★   ★   ★

This book is about the evolution of dog-diagnostic behavior characteristics. It is about the origins of breed behaviors, and the various symbiotic relationships dogs have with people. It shows how, and why, dogs have adapted to many human environments. Our method is to look at nine different kinds of dogs: village, livestock-guarding, hound, sled-pulling, herding, retrieving, pointing, household, and assistance. The behavior of each kind is genetically adapted in a different, specialized way to living with people.

In Part I, we follow the evolution of dogs from their wolf antecedents to their commensal beginnings as domesticated village dogs, and then to the formation of the "natural" breeds.

In Part II, we examine specialized types of sheepdogs, sled dogs, and hunting dogs. We explore their unusual physical and behavioral conformations. We look at the environments in which they acquire their unique skills. These are dogs that are considered to exist in mutually beneficial relationships with human beings.

And in Part III, we explore the worlds of pet (household) and service (assistance) dogs. We ponder their present status, and give evidence that their relationships with people are amensal or parasitic, or worse. We make a case that dogs are not irrefutably man's best friend, nor is man totally dog's best friend.

Part IV looks at the most recent controversies about dogs—about their genetic relationship with other canids, their scientific name, their age as a species—if they are a species—and finally how they are able to change shape so rapidly into so many diverse forms.

Our exploration of these questions should lead you to an even greater appreciation of your dog than you already have. You will be able to look at dogs in a new way, and recognize details in your dog's behavior that you've never noticed before. You'll probably view all dogs with a new respect. If you work with dogs, as a pet owner, trainer, handler, veterinarian, or are in any way connected with dogs, we believe this book will give you new insights into your dogs that will help you work much more effectively with them. And that will be fun not only for you, but for the dogs.

PART I

THE EVOLUTION OF THE BASIC DOG:
COMMENSALISM

Every place I go in the world, there are village dogs. They feed in streets, backyards, and dumps. They tend to be smallish but uniform in size and appearance—less than thirty pounds, seventeen to eighteen inches at the shoulder, with smooth, short coats and "tulip" ears; any color or colors are "standard." I'm seldom surprised by their presence or what they look like. They don't seem to belong to anyone or need anyone. They evidently depend on the village but humans hardly notice them. They are enjoying what ecologists call a commensal relationship. They feed at the same table (so to speak) with people, but were they invited?

I believe that what I am seeing is populations of dogs that have existed—almost unchanged—for thousands of years. They are ubiquitous, and so nearly identical in size, shape, color, and behavior that they look to me like an animal that has achieved its perfect state for survival in its niche. They feed, reproduce, and avoid injury in a one-sided relationship with people in villages.

I am looking at contemporary descendants of the original dogs. I think they are modern examples of the canid "missing link," a protodog, if you will, the first canids with the traits of domestic dogs.

CHAPTER 1

Wolves Evolve into Dogs

A T SOME POINT in human history there were no dogs. There were wolves, jackals, and coyotes, but no dogs. Something had to happen in order for the wilderness creatures to transform into the domestic dog. Some evolutionary process had to take place. But what could it have been? Was it artificial selection by people, where people took wolf pups from dens and tamed them? Or was it natural selection, where a changing environment dictated survival of the fittest? If it was natural selection, then the wild animals had to be adapting to a new, or changing, niche. And that new niche had to have something to do with people, because the result was a domestic ("housed") animal.

Although dogs resemble wolves superficially, when I look at dogs I never think *wolf.* In form and behavior there are huge, important differences. Wolves are wild, they live in the wilderness, they avoid people, and they kill their own food, often in cooperation with other wolves. Dogs, in contrast, live around human habitations, and rather than avoiding people, look to them to provide food. A basic change, a *genetic* change, has occurred. This difference in food-gathering behavior has led to an animal that is tameable and trainable, and these are genetic traits.

Darwin published his great work *The Origin of Species* in 1859. He revealed to the world his carefully studied, well-constructed rationale about natural selection, often contrasting this with examples of artificial selection. In marveling at the variety of forms and colors within a domesticated species, he, and most other students since then, thought that selection for dogs, for example, was by artificial selection. People were the selective agent for the transformation of wolves to dogs. They imagined people taking wolf pups from dens and taming them and then training them to be useful. The theory was that wolves and people

This wolf cautiously picks up scraps we threw to it on the highway in Quebec. Many have hypothesized that this was the first step in domestication. But I doubt that throwing scraps to an unusually tame wolf is even remotely close to selective breeding.

Stan Warner tries to prevent this fox from stealing our fish. Of course, we taught her to steal fish so we might get a close-up picture. We also taught this little girl to lick our dishes clean. It is easier to teach wild animals to come to a spot with food than to get them to take you to food.

teamed up for their mutual benefit. Each offered the other something. Perhaps humans provided the brains, and wolves the speed and killing skills.

But there's a logical fallacy here. Just because some people select and breed dogs now does not mean that the original dogs were created that way. Looking closely at the behavior of wolves, and understanding the biology of a wild animal, I don't think there is a ghost of a chance that people tamed and trained wild wolves and turned them into dogs. I think a population (at least one) of wolves domesticated themselves.

THE PINOCCHIO HYPOTHESIS OF DOG ORIGIN

The widely popular view is that people created dogs by artificial selection. People took pups from wolf dens and made pets out of them. They tamed them, trained them, and took them out hunting. After many generations of this regime, the wolves evolved into dogs.

The biological implausibility of this leads me to flights of fancy, and I tend to call it the Pinocchio Hypothesis. The old wood carver, wishing he had a son, carved himself a wooden one and named him Pinocchio. The lad wanted nothing more than to truly please his father by becoming a real, flesh-and-blood boy. He wanted this so much that, after learning a few important lessons about behaving like an honorable person, he qualified for a fairy godmother, who waved her wand and turned him into a real boy. Changing wolves into dogs by getting them to behave like dogs has the same kind of ring to it.

What the wolf-pup-into-dog hypothesis is really depending on is that wolves are related to dogs (true), that dogs easily form associations with people (true), and that therefore wolves must have formed relations with people in the past (not true). Other vague assumptions are necessary to support the hypothesis. For example, tame wolves supposedly behave more like dogs than wild wolves (also not true).

Once the wolves are tamed, can they be trained to do something useful, perhaps barking when strangers come, or helping with the hunt?

According to the Pinocchio Hypothesis, once the wolf is tamed and trained to human wishes, it turns into a dog. Somehow Mesolithic people, who had never seen a domestic dog, selected among those tamed and trained wolves for doglike characters that no wolf has ever shown: floppy ears, spotted coats, and other characters we think of as strictly doglike. For this to work, one must assume that there were enough of these tamed and trained wolves to sort among, and breed the best to the best. Over a span of many years (hundreds? thousands?), the Pinocchio progression looks like this:

(Learn) Tame — (Learn) Trainable — (Genetically) Domestic

First the wolves learn to be tame, then they learn to be trainable, and the eventual result is that they become (genetically) domestic dogs. I'm not sure just how the genetic mechanism is supposed to work here, because no matter how well tamed an adult wolf might be, it will never give birth to genetically tame puppies.

Wolves are not an animal I would portray as (very) tameable or (very) trainable. Wolves can be *taught* to exhibit a few tame qualities (unfearful of humans, and tractable to a very limited extent). However, dogs are *genetically* tameable and *genetically* trainable. Dogs *are* tame. And that is a huge genetic difference.

Let's test these assumptions of "tameable" and "trainable" and decide whether the Pinocchio Hypothesis is plausible.

TAMING THE WOLF

The first question is, Do "tame" wolves act like dogs (or, Can wolves be truly tamed)? Wolf biologist Erik Zimen in Germany found that when he attempted to socialize his captive wolf pups after they had reached about nineteen days, he never succeeded. Compare that with a dog, which still has the ability to be socialized with humans at ten weeks of age. At nineteen days, dog and wolf pups are still dependent on milk and won't begin to eat solids for another two weeks. At captive-wolf locations, such as Wolf Park in Indiana, wolf pups are raised from the age of eight to ten days by human "puppy parents," the goal being to achieve adult wolves that are accustomed to human care.

Psychologist and animal behaviorist Erich Klinghammer, director of Wolf Park, is one of the world's experts in taming wolves. For years, he and his colleague Patricia Goodmann, and a host of specially trained wolf-puppy raisers, have "socialized" the pups with people. Hand-raising is an elaborate, laborious procedure, but absolutely necessary. The performance of the day-to-day routine of entering cages to clean, feed, and otherwise care for the wolves, requires that the wolves be unafraid and accepting of people. At Wolf Park, wolves are also habituated to being attached to a leash, so they can be easily moved from place to place. Now, try to imagine people in Mesolithic villages performing these socialization processes. It takes a great deal of dedication, and what would be the survival value for ancient peoples?

No one at Wolf Park—least of all the handlers—is ever fooled into believing the adult wolves are pets. In fact, handlers operate as if they are entering the wolves' world, and they behave within the social rules of wolves. They are carefully taught how to work with the wolves without antagonistic confrontations. The handler does not try to behave like a pack leader and make the wolves submissive. Klinghammer believes the wolves view their familiar handlers as though they were other pack members. He has found that by behaving positively with the wolves, in ways that are natural (innate) to them, and thus triggering the release of

Socializing a wolf pup. Wolf Park has dedicated personnel who take wolf pups from the den at thirteen days old and spend the next months with them. It is absolutely necessary to start before their eyes are open if they are to imprint on people. (Photo by Karin Bloch)

their innate social behavior, he can get the response he wants from them. For example, in trying to make contact with shy wolves, handlers have a better chance if they wait for the wolves' usual rally among themselves, at dusk and dawn. Because the wolves are focused on their rally, the handler is more likely to be able to greet and groom the shy one. But even with all the advanced knowledge and techniques in use now, when all is said and done, the wolves are only partly tame, and they are still dangerous.

In fact, a wolf that is unafraid of humans is more dangerous to humans than is a wild wolf. A wild wolf will flee when approached, but a "tamed" one is not afraid to move in and bite. Twenty years ago, Erich took a reluctant me into the main cage with the "socialized" wolf pack. All wolves had been born in captivity for several generations, hand-raised as puppies, and "tamed." All were part of Wolf Park's demonstrations and were handled daily. Why was I reluctant?

"Look," I said, "I know about dogs, and I have observed wolves in the wild, but I don't know much about tamed wolves."

"Just treat them like dogs," Erich told me authoritatively. Which I did, by thumping Cassi on her side and saying something like "good wolf." That was when she became all teeth. Not a nip, but a full war—a test of my ability to stay on my feet and respond to Erich's excited command, *"Get out, get out!* They'll *kill* you!" Note the wording "They'll *kill* you!" I had a blurred vision of a collection of wolves gathering, and a wolf tugging on my pants as Cassi focused on my left arm.

"Why did you hit her?" Erich said later, almost too softly to be heard over my pounding heart.

"It wasn't hitting! I was patting her! You said treat them like dogs and I pat dogs and if I do some social misconduct with a dog, I don't get my head bit off, and why is it that all you people who socialize wolves have those nasty scars!" I said in a single breath while applying a tourniquet to the mangled arm of my goose-down jacket. Never again did I think that tame wolves could be treated like dogs. At that point, Erich and I began to seriously explore differences in behavior between dogs and wolves.

Everybody I know who tames wolves has equipment. The Pinocchio system of taming wolves entails a large canid management program. First of all, it takes considerable ingenuity just to contain a reproductive wolf. Tame wolves don't exactly sit around on the front porch and wait for you to come home. I'm not trying to belittle the Mesolithic imagination, but keeping wolves from running off is not easy. Those long-ago people would have needed the Mesolithic equivalent of fences, collars, and chains. In his book *Man, Culture and Animals,* anthropologist M. J. Meggitt reports on how the native people would take dingo pups from the den and tame them, and the tamed dingoes would hang around the village for a couple of years. But then came the call of sexual readiness, and they were gone. The aborigines knew this pattern, and women who took a fancy to a particular dingo would break its front legs so it couldn't return to the wild. Now there is a management strategy that might have worked on Mesolithic wolves, but a wolf thus afflicted wouldn't have been much good at hunting.

Can a wolf be tamed so that it forgets its wild ancestry? And if so, does domestication follow easily? Motion-picture photographers Lois and Herb Crisler spent two summers and a winter in northern Alaska during the 1950s. Each summer they raised a litter of captured wolf pups so they could photograph them. In her book *Arctic Wild,* Lois Crisler describes life on the tundra with highly personable wolves coming into the camp and going out at will, scenes that certainly dispose one to believe that early people indeed could have "tamed" wild wolves. The young wolves seemed to accept the two humans, greeting them, playing with them, interacting in ways that were "almost human." But note the word "young." Many wild animals can be tamed and played with when they are young. But most revert to wild instincts and behavior as they mature.

To equate the Crislers' experience with that of early man, and to hypothesize that domestic dogs could result from a few people-friendly wolves, is, again, to overlook basic principles of evolution. "Tame" the Crislers' wolves might have been, in that they shared food and shelter with them, but at reproduction time off they went, back to the wild. They were not truly tame. The wolves figured out how to get out of their pen, even to the point of knowing on which side the fence was weakest. They were allowed to come and go as they pleased, but they did it on their own schedule. The Crislers were motivated to accept the wolves on the wolves' terms, in order to preserve their wolf behavior for the motion picture camera. They did not ask the wolves to help with hunting or personal protection, although the Crislers obviously enjoyed their company both in the field and in their small cabin.

Even the Crislers, as highly motivated as they were to keep their wolves later on as pets, realized they were not in fact tame—not in the sense that dogs are tame. Recently, as part of a documentary film, I interviewed a number of people who kept wolf hybrids. Most of them had bought a dog that was purported to be one-eighth wolf, or some such unconfirmable combination. But several had true hybrids. The hybrids had been bred purposely at a zoo, or as part of an experiment, and one woman had an accidental litter with her hand-raised wolf and a German shepherd dog. Every one of them said, "It was a terrible mistake. I'd never do it again." They raised these animals, loved them, and stuck with them to the end. The stories they had to tell were numerous, because every day was an adventure. Their stories were also hilarious (in retrospect). Each had built an escapeproof cage, with double sliding doors, and each told about the time the animal got out of the cage and ate all the ducks, or the cat. I did go into the cages, but I didn't pat them. Everyone told me to "be careful, don't make any sudden moves." I did feed the one who carefully took dog yummies from between my fingers. But still, those wolf-dogs were not my idea of canine companions. They were nothing I would want for a household pet, nor did their owners ever want one again.

All of us who have kept wolves and coyotes in kennels realize they are masters of the escape. I have kept dogs in kennel runs for years without their solving the latch problem. Of course, we have had exceptions (there are always exceptions). Some of the border collies figured it out, and old Tom the Maremmano-Abruzzese could get a chain-link gate open if the latch was the right kind. But wolves and coyotes figure it out the first day. Not only that, they then open a friend's cage and all the other cages. In our kennel is a New Guinea singing dog, a dingolike canid that authorities believe is still semiwild. Caruso "knows" that if we accidentally leave the snap hook off the latch, he can get out. But he is too smart to let *us* know he can get out. He doesn't flip the latch because he also "knows" he'll just be trapped in the kennel. To really escape, he must wait until not only the kennel door snap hook is missing, but the outside safety door is also open. Sooner or later every new kennel worker makes this mistake, and when all the stars line up, Caruso is gone—usually on a beeline to our flock of sheep.

In other words, these wild forms show different cognitive abilities. Dogs are not insightful in the same way the wild types are. Wolves and dogs learn differently. This has been demonstrated experimentally by psychologist Harry Frank and his colleagues, who studied wolves and dogs (malamutes) during the 1980s. They suggested that canid problem-solving strategies were different. Wolves seemed to be learning by insight, but dogs needed repetition in order to learn a task. Researchers at the universities of Connecticut and Michigan reported that their wolves were much better at solving problems, especially those involving sequential manipulations, than were their dogs. The Crislers in Alaska observed that their wolves would watch intently while they slid open the bolt that kept the door closed and then try to repeat the process themselves (often successfully). Dog trainers have the opposite experience—dogs are in general unable to learn simply from demonstrations.

One semester we were raising coyote pups with border collies of the same age and sex, in order to observe developmental differences. Each week we weighed and measured their progress. The coyotes were shy and hard to weigh, especially when the border collies were bouncing all over us and the scales. One particularly disruptive day we took the coyotes to an empty kennel. While we weighed the first one, the other climbed up the side of the chain-link cage and walked along the top rail back to his home cage with the collies.

For the first time I realized they could get out any time they wanted to. "Could get out" is a euphemism for "were getting out." To gauge the extent of this escape artistry, I adopted the wildlife manager's technique of spreading a little flour on the floors and around the outside of the building. The next day showed telltale footprints. Every night they were out in the field by the kennel hunting mice. Why didn't the border collies watch and learn how to get out? Why did the coyotes come back? Good questions.

The important point about these anecdotes is that not only were the wild animals better at solving a problem, but they learned how to solve a problem by watching another animal, including humans. Dogs as a rule are very poor at observational learning. NB · Opposing view/ opp theory.

TRAINING THE WOLF

Now let's look at the other hazy assumption about wolves, that once tamed, they can be trained to do something useful. We have seen that they can learn to form relationships with people, although not easily, and that the relationships work best when they are on the wolves' terms. But then can they be trained to some task? What about shaking hands or walking to heel on a leash? At Wolf Park, although the wolves are taught to accept being on a leash, they are not asked to walk at heel. They are free to walk (and sometimes run, with the trainer) within the length of the leash.

I have never found tame wolves responsive even to modest suggestions, including simple ones like "sit" or "stay." "Come" also is not a big winner among wolves. Wolves are not exhibited at obedience shows. I have never seen circus performances with wolves, jackals, or coyotes. People can train lions, tigers, and leopards to sit up and jump through flaming hoops, but with all the interest in wolves, why hasn't anyone taught a wolf to sit on a chair with a wolf biscuit perched on its nose? And balancing a biscuit on the nose isn't very complicated when I think of what dogs can be trained to do: herd (and guard!) sheep, point pheasants, retrieve ducks. Now someone will send me a photo of his wolf pointing pheasants, but before I am convinced, I need 1) to see the wolf do that over and over, and 2) to locate a brochure from the Wolf Retriever Club of America.

How about training wolves to pull a sled? This task is done commonly by Siberian huskies, a breed of dog that people like to identify as being "closely related to wolves." I guess the assumption goes: if huskies are so good at it, and are closely related to wolves, then the wolf should impart *some* advantage to a sled team. In the most serious trial of this that I know of, zoologist and wolf expert Erik Zimen trained a team of wolves to pull a dog sled. In his book about the wolf is a picture of a team of sled wolves. They look great in the picture, but the story belies the apparent serenity.

Zimen worked with five zoo-born, hand-raised, nearly year-old wolves, training almost every day during an extra-cold, long, snowy winter. Eventually they all accepted the harness, the weight of the sled, and the need to travel in a straight line. But they were ineffective and unreliable as sled pullers. They tended to be more interested in preserving their own personal space, and this was dicey with them hitched together in harness. "Gee" and "haw" didn't work; they chose their own direction. One or two would decide they had gone far enough and just lie down. Zimen's description ends with a story of a robin intruding among the resting team. The wolves climbed over each other to sniff, chains got tangled, the wolves got increasingly aggressive with each other, and a monumental altercation ensued. The wolves were finally controlled, but Zimen and his wife tugged the sled home that afternoon.

I raced sled dog teams for over a dozen years, and can attest vigorously that the problems already visited on the driver even by a team of good working sled dogs need no exacerbation due to the addition of a wolf. People who sell huskies often claim "wolf blood," for it excites the novice into purchase. Perhaps it really just helps to sell pups that have no purebred pedigree and cannot be registered as purebreds.

But what possible wolf trait would anyone want to add to a sled dog team? Wolves are not faster or stronger, nor do they have more stamina than sled dogs. They do not take commands better, if they take them at all. I have never heard of even one wolf that has demonstrated top racing or freight-pulling ability. The very important distinction between dogs and wolves is that wolves may be smarter and able to learn more, but they cannot be taught much using instrumental conditioning. Dogs have been described as just smart enough to do a job and just dumb enough to do it. They are without peer in the animal world in responding positively to instrumental conditioning.

DOMESTICATING THE WOLF

Even if Mesolithic people were able to tame generations of wolves, and then train some of them to obey simple commands (for example, to come when called, or sit), still the burning question remains: What changes wolf genes into dog genes? Taming and training might change an individual wolf slightly, but they do not change its genes. We still need that fairy godmother to turn them into dogs. Unless we believe in fairy tales or Lamarckian inheritance (the inheritance of acquired characteristics), we have to conclude that taming and training are not enough.

The biological reason that belies the Pinocchio Hypothesis is that the offspring of tamed (and/or trained) wolves do not inherit this tameness. When tamed wolves reproduce, they get wild pups (oriented away from human activity). When dogs reproduce, they get tame pups (oriented toward human activity). The two species are intrinsically, instinctively, genetically distinct, one from the other, in this respect.

A concise statement about this biological dichotomy was written by Australian wildlife biologist Laurie Corbett in his 1995 book, *The Dingo in Australia and Asia*:

> In a practical sense, the progeny of a long line of domesticated animals will "automatically" perform the activities required by its human owner, whereas the tamed progeny of a wild animal requires retraining (retaming) with every generation. Thus, it is theoretically impossible to domesticate a wild animal and maintain its natural behaviour patterns, because two genetic makeups are involved and an animal has to follow one pattern or the other.

The Pinocchio Hypothesis, then, does not fit the behavioral facts. The biological problem here is that tamed and trained wolves cannot pass those learned tricks to their offspring. To postulate artificial selection among wild wolves by Mesolithic people, starting with puppies from the wolf dens, with the goal of producing domestic dogs, is wishful thinking.

Since wolves can't pass their tame and trained characters on to the next generation, then we have to postulate some other mechanism for the transformation of wolves into dogs.

SPECIATION REQUIRES POPULATIONS THAT EVOLVE—NOT INDIVIDUALS

Selective breeding is the way most scientists have visualized the creation of dogs from wolves. Since it is *populations* of organisms that evolve, not individuals, speciation would require that many many Mesolithic peoples tamed and trained many many wolves. Some wolves were easier to tame and train than others. These easier tamed wolves tended to breed with other wolves that were easier to tame. In addition, Mesolithic peoples noticed that in each generation of wolves, some were more lovable than others, and so they bred these wolves to other lovable wolves. Eventually, they called the resulting strain of lovable wolf, dogs.

One axiom of speciation is that the frequency of some (genetic) character in a population of animals changes over the generations. Lovable in the ancestral wolf population exists at a very low frequency.

Through selection, the frequency of lovable increases over the generations, and the resulting new "species" differs from its wild ancestor in at least this one character. If this lovable trait isolates this population sexually from the ancestral population, then one could say that speciation has occurred.

A new species, then, evolves through the gradual shifting, over time, of gene frequency within a population.

When evolutionist Charles Darwin explained his theory of natural selection, he used the analogy of farmers creating new breeds by *sorting* among their herds, *selecting* individual animals that showed certain desirable traits, and then *breeding* these animals. An important part of this concept is the fact that farmers are sorting through a herd, a population of animals, to develop a new strain. They are not creating cows, but selecting among the best milk producers, say, in order to produce a strain of superior milk-producing cow.

Breeding fancy pigeons was great sport among the British in the mid-nineteenth century, and Darwin's neighbor was a pigeon fancier. Darwin spent hours among the pigeons, observing how his neighbor sorted among the flocks (a population) of pigeons with exotic forms or behavior, in order to create various new races. Darwin called this "artificial selection," that is, selection by people rather than by nature. But here again, what's important is that these pigeon breeders were selecting individuals that varied slightly in visible traits from other members of the population. They were not reaching into a cage and taking a bird at random.

I think this is my biggest problem with the Pinocchio Hypothesis. There is no archaeological evidence that Mesolithic people had a big enough population of trained or tamed wolves living among them. In order to select for a tame behavior or a doglike shape, people would have to be able to *see* differences in wolf behavior and wolf shape in the existing wolves. How did Mesolithic villagers sort among wolves for the genetically tamer ones? In order for wolves to go from "learned" tame to "genetic" tame, a Mesolithic breeder would have had to sort through hundreds of wolves, find variations in natural tameness, isolate those tamer wolves from the rest of the population, and breed only the most tame, for many generations.

Darwin thought that artificial selection mimics natural selection. Farmers achieve production goals by selective breeding, for example, for more eggs or milk from their domestic stock. Darwin was so fascinated by what people could do by artificially selecting that he imagined a hypothetical being who would have the power of natural selection and "could study the whole internal organization . . . and go on selecting for one object during millions of generations . . . who will say what he might not effect?"

Darwin thought that evolution took place very slowly. The changing of form from one species to the next took place increment by tiny increment, generation after generation. In order for his natural selection model to work he had to postulate unlimited generations. But there is no such time span for the evolution of domestic dogs. There might—possibly—have been dogs fifteen thousand years ago. There's lots of evidence that breeds of dogs existed by 8000 B.P. So, in about four thousand years, we go from not being able to tell the difference between wolves and dogs to having dogs all over the place. For an evolutionary biologist this is just too much to expect from natural selection. Transformation from wolves to dogs could not have happened by gradualism in such a short period of time.

Darwin and those who came after him simply assumed that people domesticated the "dogs" by sorting through the variation in the wolf population. Darwin assumed that since people created breeds of dogs during his time, they must have done it earlier.

SPECIATION REQUIRES DIFFERENTIAL MORTALITY

Another concept underlies speciation, whether by artificial or natural selection. This one depends on the second assumption of evolution, that development of a new species requires differential mortality. Differential mortality in artificial selection means that the farmer allows some animals to breed and prevents others from doing so. The ones that are prevented from breeding are genetically "dead." In the wild, says Darwin, animals will vary (genetically) in each character (at least slightly), and the animals with the fittest (genetic) characteristics will produce more and better offspring than those that go around tearing the arms off goose-down jackets. Again, in order to have a differential mortality, we have to postulate a population where tameable and trainable individuals leave more offspring than the wild ones.

Darwin wrestled with this problem of where the variation came from that people were selecting. In order to get around the lack of variation in wolves, Darwin hypothesized that dogs descended from multiple ancestors—including both wolves and jackals. This hybridization would give the variation necessary for selection to take place. Contained in his theory of natural selection (the survival of the fittest), and also in artificial selection, is provision for genetic variations in all the traits that make up an individual. Without these variations, there could be no differential mortality. Darwin realized that for many of the evolved characters of the dog—for example, the two estrous cycles a year—there was no variation in wolves. How on earth, then, could you select for it? Darwin didn't have access to the information we have now, and some of us think he would have been flabbergasted by how little variation in wild species there is over time. There are fossil records of species that enter the record, live for twenty million years, and go extinct without changing a bit. Indeed, the Canidae have been around for forty million years, and they still have identical skull-length proportions. They still have weak jaws and lousy eyes, compared with other predatory species. In forty million years, the best they can do is to change size.

When we begin to play by the rules of this Darwinian game, we begin to realize that the early domesticators would have had a real problem. They needed an accessible population of wolves—a large population—and it needed to show variation. Trying to pick out the genetic components of tameable and trainable would not have been easy. Then they had to breed these animals—which had minds of their own about breeding. Add the fact that wolves don't come into heat until their second year, and Mesolithic wolf breeders would have had at least a four-year wait until they could have sorted and bred their first born-in-captivity wolf pups. Would any significant changes have been evident in this generation? It's doubtful. Five generations of arranged wolf breedings might have been about it for each generation of people. Delayed gratification, then, and long-range planning, were musts for the early domesticators of wolves if the Pinocchio Hypothesis is correct. Even then, success is questionable. Ethologists at Wolf Park have been socializing wolves for twenty-five years, generation after generation, and the wolves have not turned into dogs. (They haven't been selecting for tameness, but simply socializing the pups.)

In the meantime, if they had any intention of making dogs out of wolves, Mesolithic people must have been selecting for a host of other characteristics: brightly colored, variegated coats, floppy ears, stick-up tails, and a six-month estrous cycle starting at seven months instead of two years. And don't forget barking—wolves are not great barkers. In fact, wolves rarely bark. Think of selecting among wolves that don't bark at all for those that bark only rarely.

What is a real puzzle with the artificial selection model is that all those traits that distinguish dogs from wolves are not variations in wolves. No variation means no selection.

It seems to me that most dog characteristics are impossible to explain by gradualism. Think of wolf tails going up or ears coming down, millimeter by millimeter, over the centuries. Mutations like the short legs of a dachshund could not be products of gradualism. They do not get shorter and shorter over generations.

That leads the artificial selectionist to postulate mutations. Why not mutations? Why couldn't the boundary between wolves and dogs be characterized by a high mutation rate? Early domesticators could have been very alert and selected those individuals that were genetically altered by mutation, and then bred them. Well, again we would face the problem of population numbers, as well as presuming that Mesolithic people could have controlled the breeding of their wolves. Arranging breedings between wild animals is not simple.

Another problem with postulating mutations is that mutations tend to be rare in nature, and most of them are deleterious. Early domesticators would have had to be very lucky to get the right ones at the right time.

The last problem with the Darwin artificial selection model is that dogs differ from wolves in some significant ways—and few people even now are aware of them. If we compare a hundred-pound dog with a hundred-pound wolf, the dog's head is 20 percent smaller. That is a *lot* smaller. Wolves have impressively large heads.

A dog skull that is exactly the same size as a wolf skull requires the dog to be a big one, weighing 150 to 180 pounds, and the wolf to be about normal, one hundred pounds. Yet even though their skulls would be the same size, the dog's brain is still 10 percent smaller. In a big dog's little head is, relatively speaking, a tiny brain. A hundred-and-fifty-pound Saint Bernard and an eighty-pound wolf have the same size head, but the wolf's brain is bigger.

Saint Bernards might have been a poor example for this picture, because they actually have smaller heads than other dogs of the same weight. The heads of Saint Bernards are surprisingly small. Their apparent big-headedness is a consequence of their very thick skin, a trait that leads in part to the saggy skin of their cheeks and around their eyes. Dog skin is generally thicker than wolf skin, but in some breeds, like the Saints, it is extremely thick. Eskimos tell us that dog-skin pants far outlast wolf-skin pants, and dog skin doesn't tear when you sew it. And dog-puppy skins are preferred for children's garments, being soft but tough.

Dog teeth are also smaller than wolf teeth. Dogs have the same

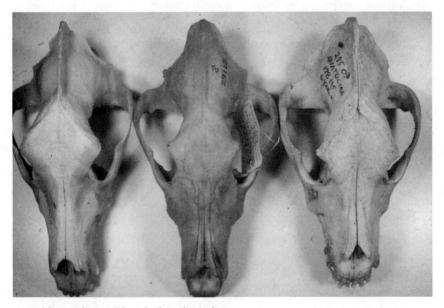

A wolfhound, a wolf, and a big sheepdog.

	Wolfhound	Wolf	Anatolian
Body weight (lbs.)	150	100	180
Skull size (cm²)	348	352	343
Brain volume (cm³)	303	329	312

Even though the dogs are at least 50 percent bigger than the wolf, they have heads that are about the same size, with smaller brains. If we were to look at the teeth on these three animals, we would find that the wolf's teeth are bigger and more robust, even though its skull is the same size and overall it is a much smaller animal.

number and kinds of teeth as wolves, coyotes, and jackals, but even twenty-five-pound coyotes can have bigger, more robust teeth than dogs of twice their size. In comparison with those of the rest of the genus *Canis,* dog teeth are cute.

Now why would the Mesolithic Dog Domesticators' Club sort among wolves for tiny brains, gracile teeth, small but bony skulls, and thick skins? Would they even have known? It is also curious why, if early domesticators were trying to develop hunting companions, they would be selecting for animals that had such poor hunting equipment. What was the benefit of having small teeth, such a little head, a weak bite, and a less intuitive brain?

Something important is missing from the Pinocchio Hypothesis. I think there are too many assumptions that cannot be supported by biological facts and evolutionary theory. There is no solid evidence that people domesticated wolves by artificial selection, sorting through variations produced by mutations, or selecting for variations that were not visible.

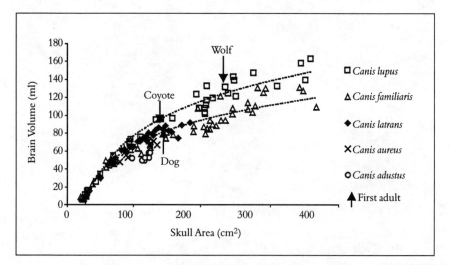

Changes in brain volume and skull size during growth. Dogs, wolves, coyotes, and jackals all (almost all) have the same head size and shape and brain size when they are born. They tend to have the same growth trajectories until they are ten or twelve weeks old. Wolves grow big heads and big brains. Big dogs have the same growth trajectories as wolves, but their brain growth slows and their brains don't get much bigger than those of four-month-old wolf puppies. (Adapted from an unpublished thesis at Hampshire College by student Ryan Kerney)

Thus far in this chapter, I have outlined a potential scenario for the evolution of wolves into dogs by artificial selection. In brief, it goes:

Capture a wolf.
Tame the wolf.
Train the wolf.
Breed the wolf to other tame, trained wolves.
And, presto! a domesticated dog.

Or, as I stated earlier in this chapter,

(Learn) Tame — (Learn) Trainable — (Genetically) Domestic

or,

Adopt — Socialize — Train — Select.

I believe this is backward. I believe the model of dog-from-wolf genesis is:

Domestic — (Genetic) Tame — (Genetic) Trainable

or,

Domestic — Naturally Tame — Naturally Trainable.

The outline goes:

People create a new niche, the village.
Some wolves invade the new niche and gain access to a new food source.
Those wolves that can use the new niche are genetically predisposed to
    show less "flight distance" than those that don't.
Those "tamer" wolves gain selective advantage in the new niche over the
    wilder ones.

In this model, dogs evolved by natural selection. The only thing people had to do with it was to establish the villages with their attendant resources of food, safety, and more opportunities for reproduction, which provided the naturally tamer wolves with increased chances of survival.

What, exactly, is this new environmental niche these animals are adapting to, and why are small heads, teeth, and brains an adaptive advantage in that niche? What I have to do is demonstrate a model for a natural evolution of wolves to dogs that accounts for all the features of the dog. At the same time, the model must fit with what we know about natural selection. It must contain more plausible mechanisms than does the artificial selection hypothesis.

In the late summer of 1998 I visited Down House in County Kent in England, where Charles Darwin raised his family and wrote *The Origin of Species.* His library is there, and his numerous laboratory tools, just as he left them. As I stepped quietly about, reading the titles of the books on his shelves and trying to absorb the flavor of his mid-nineteenth-century life, I began a conversation in my head with Darwin, about dogs. I'd like to visit again, but this time I'd like Darwin to be there, because I think we would have a marvelous discussion while I try to convince him that dogs are primarily a product of natural selection.

Mr. Darwin, I'd say, I think you would be excited by a different hypothesis for the evolution of dogs. I would like to suggest to you that the transformation of wolves into dogs happened rapidly, in just a few generations, not gradually over many generations. This is not to negate at all your great theory, but to introduce you to new evidence we have now that supports this. The evidence also accounts for one of the puzzles that you couldn't explain, what you termed "the mysterious laws of correlation."

Charles Darwin's natural selection theory is based on the assumption that the environment changes, slowly, and animals in that environment, in order to survive, must adapt to the new, changing niche. In the natural selection scenario, a population of wolves would adapt to the new niche created by people. That niche is the village. At the end of the last Ice Age, approximately fifteen thousand years ago, coincidental with tremendous worldwide changes in climate, people began to gather for the first time in permanent settlements. The appearance of villages is fairly rapid, and coincidental with the first fossil evidence of dogs as we know them.

Why is this niche important? A niche is a place where energy exists. "Energy" in this context is another word for food. Food is organic nitrogen for making protein, and calories for energy. A species might have other requirements such as a place to den, or a lack of predators, but the most important attribute of any niche is a *steady* supply of food.

In nature, animals compete intensely for food. That is because animals are able to produce many more offspring than sunlight can create food to feed them. Most of those born will die of the effects of starvation. A stable wolf population loses 70 to 90 percent of its puppy crop. Although a female wolf might produce twenty-five pups over her lifetime, only two need to grow up to replace her and her mate. That means 92 percent must die if the population is to be stable. This is a normal mortality figure among carnivores.

Is survival of those two pups random, or does it depend on something? Darwin pointed out that in every population, individuals vary slightly in every character. They vary not only in size and shape but in performance. These variations give one individual a slight advantage over another in performing in a changing environment. Some will be more suitable or more efficient in the new environment than others, and they will then leave more of their offspring to the next generation. These offspring inherit those traits of their parents, and then in turn pass them on to their offspring.

When people gather in one place for a long time, discards of waste are created: inedibles like bones and pieces of carcass, seeds and grains, rotten greens or fruits. Sometimes there will be a small surplus of uneaten food, and always the wastes of human digestion.

For simplicity, let's call this new niche the town dump. Even today, dumps are attractive places for many wild species, including wolves, bears, eagles, gulls, rats, cockroaches, bacteria, molds, and fungi. Wild wolves are common in dumps, a fact that many a wildlife photographer has made use of by caching himself with a camera near a dump and catching wildlife up close—while carefully excluding the detritus from the frame.

My colleague Luigi Boitani, a restoration ecologist, has studied wolves feeding in Italian dumps. Mountain villagers rarely see the secretive wolves. They think the wolves come to the dumps only in winter, because that is when they see their tracks in the snow. But the wolves are there feeding all year long, at night, when the people are asleep. To prove it, Luigi turned on the spotlights one midnight, and there were a dog, a cat, and a wolf with a mouthful of day-old spaghetti.

Modern wolves take chances when they approach human habitation. People are hazardous to their health. People persecute wolves, because wolves commit the unforgivable sin—they kill domestic livestock. In Italy, natural wolf food is sparse and sheep flocks are protected from predation by shepherds and guardian dogs (more about these in Part II). Boitani feels that dumps are an important food source for the wolves.

But, wolves are not efficient dump feeders. They are much too nervous. Nervous energy requires extra calories and more food. When humans appear, wolves snort an almost imperceptible single alarm call—the "wuff!"—and the pack evaporates, never seen by the approaching humans. The wolves flee, not to return until all is safe, maybe the next night. They are naturally shy of people, they run away (perhaps sooner than they need to), and they don't come back (until long after they could).

Therein lies the problem for dump invaders. Running away too quickly and too far from the food source uses up precious energy fast. How shy the animal and how far its retreat are a measure of a species's "flight distance." Some species will flee quickly when approached by humans or potential predators. Wild wolves have a long flight distance; they start running early, and they run rapidly and far. Consequently, the wolf needs more food energy in order to feed in a dump than some other species, especially dogs, which are not so nervous about approaching people. This is a large difference between dogs and wolves.

Why are dogs better at feeding in the dump? Partly because dogs are genetically tamer, and outcompete the genetically shy (of people) wolves. Outcompete means that if a dog and a wolf are feeding side by side in a dump, and a person approaches, the wolf would run first, and the dog would continue to eat. The dog not only would get more of the dump's resources but wouldn't use up as much energy by running away. A shy animal cannot learn to be tame. Tame is a successful adaptation to feeding in the dump.

Variation among the first wolves feeding in the first dumps would mean that some would have been genetically less nervous than others. The less-nervous wolves would have eaten more, and turned those calories into puppies rather than using them up running away. All that was being selected for was that one trait—the ability to eat in proximity to people. Many wild animals will flee their kill if people arrive or even come close. Wild cheetahs have difficulty eating while humans are watching them. After the speedy cheetah spends its energy catching a gazelle, tourists show up and snap photographs. Wild cheetahs don't like it and leave the carcass—either never to return, or, while they hide, jackals filch the undefended food. For a while cheetah populations were diminishing in photography zones, until people learned to leave them alone. Lions, on the other hand, do not care if someone watches them eat. Wolves are more like cheetahs.

It is in this one trait, flight distance, that the hypothesis for the self-domestication of dogs by natural selection rests. The wild wolf, *Canis lupus,* began to separate into populations that could make a living at the dumps and those that couldn't. Within one segment of the population, the frequency increased of those genes that resulted in tamer wolves, and that population could be said to have been evolving toward a new species.

These village-oriented canids also began to change shape. Their new shape made them more efficient at scavenging. The scavenger wolf was beginning to behave and look doglike. Besides evolving tameness, it acquired a size and shape that were specialized for scavenging—a smallish size, with a proportionally small head, smaller teeth, and just enough brain to point it in the right direction. These wolves were fast becoming adapted to the niche, and were incipient dogs.

Wild wolves need to be big, pack organized, and specialists in their tasks in order to find, capture, and kill large and formidable but elusive prey like moose, elk, and caribou. But their bigness is expensive, especially so for their brains. Brains cost many calories to grow, maintain, and use.

The dog, on the other hand, is adapted to scavenging at dumps. Dumps have a constant supply of low-quality food that doesn't have as many calories per mouthful as does game. It is better that dogs don't have to grow big heads and much better that they don't have to grow big brains. The smaller animal, the dog, is a new form adapted specifically for the new niche, the dump.

Smallish wolves that expend the least energy survive and reproduce better than do the big, nervous hunters. Here is the differential mortality. In the new niche, smaller, calmer animals have a selective advantage over the large, cunning big-game hunters.

I can feel Mr. Darwin warm to the hypothesis. Here is a concordance of characteristics (diagnostic traits) that defines the dog. These traits enable us to identify and distinguish dogs from wolves. Head, brain, jaw, and tooth proportions all distinguish dogs from wolves. When we add the tame behavior to those characters, we have the beautifully adaptive village scavenger.

How am I doing, Mr. Darwin?

Well, he might say, it is a very interesting theory, but there are still other differences between dogs and wolves that you haven't explained. When I think of dogs I think of my fox terrier, Molly, whose favorite spot is right over there, curled up on the cushion of that chair. Why, pray tell, are floppy ears, richly patterned coats of many colors, incessant barking, two estrous cycles a year, and early reproductive readiness— why are they adaptations to scavenging the village dump? How does this village-dog theory explain those diagnostic characters?

First and respectfully, sir, I want to say that if I can't find a selective advantage for each of these doggy traits, I still haven't lost the debate. These particular characters you mention are so bizarre in the natural world that scientists have always assumed they must be products of people selecting for them, perhaps capriciously. And there is no question that recent humans have selected for bizarre characters. We don't have to assume that all the traits of modern dogs were original adaptations to the dump.

As I see it, sir, the main problem with both the natural selection and the artificial selection models is that they require an explanation for *each character*. Some breeds of dogs have a black-and-white coat, or floppy ears, or diestrous cycling, and immediately a selectionist tries to explain the adaptive advantage of a black-and-white coat, floppy ears, or diestrous cycling. Each single character, in their view, has to be a product of selection. Why, indeed, do dogs begin reproductive activities in their seventh or eighth month and have two estrous cycles each year thereafter? Wolves commence these activities in their second year, and have one estrus per year. Looking at the dog with its two estrous cycles, people will respond that Mesolithics must have wanted dogs to have more puppies and thus selected for more frequent estrus. Or I could speculate that mortality rates are higher in the dumps and the dog was responding by increased fecundity. Each of these answers assumes that the character was selected for. The truth of each answer is untestable, and falls into a category of reasoning called a posteriori, reasoning from the effects to the causes, from the facts to general principles. But just because the character is there does not mean it was selected for.

Oh, wait a minute, my friend, responds Darwin. That may be the way you modern Darwinians might interpret what I said, but don't you remember me saying, ". . . if man goes on selecting, and thus augmenting, any peculiarity, he will almost certainly modify unintentionally other parts of the structure, owing to the mysterious laws of correlation"?

Indeed I do remember, sir, I reply. It was an amazing thing for you to say. In *The Origin of Species,* you generally convinced us to think of natural selection as a process of slow gradualism, not leaps. In fact, in your first paper on natural selection, you wrote *"natura non facit saltum"*— nature does not make jumps. To me, your proposing "mysterious laws of correlation" sounds like a contradiction, because the unintentional modification of other parts sounds like a saltation. Intense selection for one character causes others to leap to a new form spontaneously. What if floppy ears, diestrous cycling, and multicolored piebald coats are leaps, or saltations? What if intense selection for dump-scavenging ability modified unintentionally these other traits? For example, maybe a smaller head with smaller teeth and a weaker jaw causes the ears to droop.

Mr. Darwin, I continue, about a hundred years after you postulated the mysterious laws of correlation, a Russian geneticist named Dmitri Belyaev was starting a long-term experiment with the Russian silver fox (*Vulpes fulvus*). You would have loved this experiment. It provides us with good evidence to explain those apparent jumps. I love it because it illuminates a highly probable biological mechanism for the self-selection of wolves into dogs. Belyaev was in charge of a huge fox-fur farm in Novosibirsk, and without any intention of doing so, he produced foxes with those same doglike traits of floppy ears, diestrous cycling, and piebald coats.

He began selecting silver foxes *solely on the basis of their tamer behavior.*

Belyaev began this experiment because wild foxes are hard to manage on commercial fur farms. Even though these wild foxes had been bred in captivity for eighty years, and raised and cared for by people (tamed), they were a problem. Foxes are difficult to keep in captivity, the same way wolves or any wild animals are. They are shy of their keepers and run away from them, sometimes snarling. Or they hurt themselves by running blindly into walls, or crowding together with other panicked foxes and overheating or suffocating. It is exactly the same kinds of problems people would have had raising and taming wolves.

The behavior of nervous, flighty foxes illustrates wild type, or natural, behavior. This is exactly the same problem I outlined with the nervous, flighty wolves at the dumps.

Belyaev and his colleague, Lyudmila N. Trut, had observed a variability in defensive behavior among the captive foxes, which they believed was inheritable. If it was, they could select for it. Belyaev's initial study population of 465 foxes, randomly chosen from among thousands, reacted to people in different ways: 40 percent were aggressively fearful, 30 percent were extremely aggressive, 20 percent were fearful, and 10 percent displayed a quiet, exploratory reaction without either fear or aggression.

But, Belyaev noted, "even the nonaggressive foxes could not be handled without special precautions against biting, so that they, too, were virtually wild animals."

The difference Belyaev sought between the captive foxes was flight distance. Flight is a hazard-avoidance behavior, an essential component of a wild animal's survival. There are two measurable components to flight distance: 1) how close you can get to the animal before it attempts to flee, and 2) how far away it runs.

Belyaev selected from among the quiet, exploratory population of foxes and bred a second generation. In succeeding generations, he selected even more strictly, finally to the point that to be selected, an animal had to "willingly" approach him (a reverse flight).

After only eighteen generations, Belyaev had come up with naturally tame animals that had many of the behavioral traits of a domestic dog. They were different in several significant respects from their unselected kennel-mates: They reacted to people actively and positively. They would search for their keepers, climb on them, take food from them, sit on the windowsill looking for someone to approach, roll over to get their tummies rubbed, and let people carry them around and give them their shots. They answered to their names.

They behaved like dogs. Even more surprising, they looked like dogs. Their tails turned up at the end, like a dog's. Their coats were often piebald, their ears drooped, and the females came into heat twice a year instead of once. Belyaev noted, "They even sound like dogs."

What is important is that Belyaev did not select for any of these

A piebald fox. Thirty years ago, Dmitri Belyaev sent me several pictures of his genetically tame foxes. He had selected only for tame behavior, but he got a number of other, even undesirable, changes that he could not account for. Among them are piebald coats, floppy ears, and doglike sounds.

traits. He did not even want some of these traits. Who would want a black-and-white piebald fur coat? Most of the other traits he didn't particularly care about, and he never anticipated any of them. It didn't matter to the fur-coat industry if the foxes had floppy ears, or barked.

Mr. Darwin, these doglike traits appeared mysteriously, unpredictably, spontaneously, if you will. Would it not be fair to say, sir, that because Belyaev kept on selecting foxes for tameness for eighteen generations, "thus augmenting this peculiarity," he did "almost certainly modify unintentionally other parts of the structure, owing to the mysterious laws of correlation"?

Now, Mr. Darwin, I hear you thinking, "But this is a great example of *artificial* selection! Which animal is selected to reproduce is under the control of people. People are selecting foxes for tameness and they get the bizarre features by the mysterious laws. If people had been selecting wolves to be tameable and trainable, they also might have got richly patterned coats, floppy ears, and all the rest."

Well, sure, they could also have got richly patterned coats and floppy ears. But by "all the rest," do you mean small teeth, skulls, and brains? Belyaev never mentioned any reduction in sizes among his transformed foxes. If he had found that these dog traits were also part of the fox-to-dog saltation package, then I have to adjust the argument. I'm assuming that small body size and small heads, brains, and teeth are products of natural selection and are not saltations. If they are part of a saltation resulting from selection for tame, then I could not discriminate between natural and artificial selection.

Belyaev's work appeals to me because it provides real evidence for the transmutation of wolves into dogs by natural selection. It provides evidence that fits observable facts.

The Pinocchio Hypothesis of dog evolution requires that "tame" be a learned adaptation by individual wolves ("take a wolf pup from the den and tame it"). The natural selection theory starts with a scavenger dog, a "tame" canid that is genetically adapted to feeding in the presence of people. Dogs remain in the presence of people because that is where their food is. The village has become their niche; that is where they are adapted to be. The artificial selection theory (Pinocchio) requires that wild wolves be trained to stay around and hunt in concert with people. But their adaptation to the wilderness would constantly pull them away from the village, just as the Crislers' tamed wolves were pulled away.

The artificial selection theory requires early people to embark intentionally on a long-term wolf-taming and breeding project, which is difficult to do, and perhaps impossible to do well enough to have any evolutionary change take place. The natural selection theory doesn't require people to do anything other than live in villages.

Okay, says Mr. Darwin, I think the argument is worth exploring further. But why are you so invested in this theory that the tameable behavior which distinguishes dogs from wolves was a product of natural selection to a new niche? Why is it important to convince me that the evolution of the dump scavenger was by natural selection? What on earth difference does it make whether my dog Molly descended from the village dog population or whether she is a direct descendant of wolves? Indeed, in either case she is a direct descendant from the wolves.

Very respectfully, sir, it makes a huge difference. Remember, when

The Origin of Species was published, how you were pictured in cartoons with a monkey body, as if being descended from apes made you an ape? Today, the popular dog press seems to feel that if dogs descended from wolves, they would have wolf qualities. But the natural selection model points out that the wolf qualities are severely modified. Dogs do not think like wolves, nor do they behave like them. Books about training dogs would have us believe that dogs get their behavior directly from wolves. We are advised to act like the pack leader, the alpha male, and treat our dogs as subordinates. Since dogs came from wolves, they say, dogs should behave like wolves, think like wolves, and respond to wolflike signals.

But dogs can't think like wolves, because they do not have wolf brains. We descended from apes, but we don't behave like them and we don't think like they do. We are a much different animal than the apes in spite of our common genetic ancestry. The same is true of the dog and its ancestor.

Laurie Corbett put it best, finishing his thoughts about the basic genetic differences between wild and domestic: "Wild dingoes, therefore, can be tamed but not domesticated. Should humans determine and selectively breed certain standards and characteristics for dingoes, they will cease to be dingoes. A domesticated 'dingo' is not a dingo but just another breed of dog."

Mr. Darwin, I think it is wrong to treat our best friend like a wolf. I have trained hundreds of sled dogs and hundreds more sheepdogs. Asserting dominance over one of my favorite working dogs by pressing it onto the ground and snarling at it is preposterous. I don't want my sled dogs rolling on their backs and urinating in the air like some subordinate wolf every time I show up. I don't think a dog knows what people are talking about when they exhibit this "alpha wolf" behavior. Dogs do not understand such behaviors because the village dogs didn't have a pack structure; they were semisolitary animals. Such behavior by humans confuses them.

The biological reality of all this is that the wolf is now the distant cousin of the dog. That canid family tree split, and wolves and dogs went along their separate branches. The wolf displays specialized adaptations to the wilderness, and the dog displays specialized adaptations to domestic life. The two canid cousins are adapted to different niches, and they are very different animals because of it.

CHAPTER 2

Village Dogs

As I MOVE around the world studying working dogs, there always seems to be another population of dogs hanging around. They are part of the background. When I ask about these dogs, they are described to me as the local mutts, mongrels, pavement specials, crossbreeds. People tend to think of the mongrel as a degenerate form, a dog with mixed parentage. A common belief is that all of these dogs have had pure-bred ancestors. When we see a mutt, we try to identify what breeds are in its genealogy. It took years to dawn on me that just because a dog does not look like a purebred, it is not necessarily a crossbreed. Just because a dog is a stray, it is not necessarily a mongrel. And how do we know it is a stray, that is, a dog no longer cared for by owners? After all, it is highly unlikely that the dog evolved from wolves into one of our modern purebreds and then later degenerated into a mongrel.

Where did this idea come from that in the beginning all dogs were purebreds and everything else since then that is not pure is a mutt? It seems that every artist's conception of the "original" dog pictures a wolfy-looking dog running with hairy human hunters garbed in something like wolf skins. The idea is pervasive and through time has led to differing speculations about dog origins. Big-dog breeds must have come from big-wolf breeds. Mastiffs must have come from Chinese wolves. The original dogs were small and must have come from small wolves. The Arabian wolves are small (now) and must be the sub-species that dogs descended from. Such arguments have a little of that "original creation" philosophy to them.

It was the Belyaev study that woke me up. The transition dog, the "missing link," between wild fox and domestic fox was a piebald, floppy-eared, diestrous, tame animal, in many ways identical to the so-called

An African village dog. Neither a mongrel nor a stray, this dog doesn't belong to anyone. It's just a sweet dog watching for waste food to appear.

mongrel street dogs. Why couldn't all these street dogs in the Middle East and Africa be direct descendants of the original dogs?

What a revelation this was for me. My boyhood dog, Smoky, was small, short-haired, tulip-eared, and black. My mom got him by contributing two dollars to the dog-food fund at the Dewey and Almy chemical company in Cambridge, where my uncle Bob worked. That is where Smoky's mother scavenged meals from both the day and night shifts in the shipping room. She hung around there all the time, and obviously made a good enough living to invest in a litter of pups. As my pup grew, another uncle, Joe, and I theorized long and hard about his ancestry. I had hoped he would grow up to be a big dog, but Smoky topped out at about thirty pounds, a somewhat racy-looking, energetic little companion. Searching through picture books of dogs, Uncle Joe and I finally decided Smoky was a crossbred Italian greyhound. That didn't seem improbable at the time, and he had to be some mixture, didn't he? Plus, it gave me an impressive answer to the frequent question, "What kind of dog is he?"

But today, when I consider Smoky's provenance, I decide that I had adopted the offspring of the basic village dog, living commensally in mid-twentieth-century suburbs. I had, unknowingly, just repeated the process that had gone on since the beginning of dogs. Smoky's parents, to my adult, practiced eye, were direct descendants of the original village dogs.

What I need to support my contention is a population of dogs living in a modern village that would simulate the conditions of the early Mesolithic villages. I need a population of dogs that live in a commensal relationship with people. The idea is to apply anthropology and behavioral ecology and show that the natural selection hypothesis, plus Belyaev's results, provide a highly likely mechanism for the transition of wolf to dog.

Several decades ago it was assumed that culture, like organisms, evolved from simple to complex, and thus anthropologists studied "primitive" cultures and "primitive" peoples. A popular concept was the "noble savage." The noble savage was one who lived a pure, primitive life, uncorrupted by modern, complex society. We now realize that "primitive" cultures are just as advanced and rich as modern ones. They are just different. One complex culture evolves into another.

This concept helps in the study of the past. For example, if anthropologists want to understand Mesolithic peoples, they study modern hunting-and-gathering societies. The assumption is that these modern societies provide an approximation of what the life of a Mesolithic hunter-gatherer might have been. Scientists compare weapons, tools, and house and food remains from middens of modern hunter-gatherers with the archaeological finds from early Mesolithic villages, and make conclusions about how the recent people compare with ancient. If the present tools approximate the Mesolithic and the people still use those tools to hunt and gather, we can assume they are culturally similar.

So, to be able to see a modern version of the original dogs that had adapted to life around a Mesolithic village niche, I looked for a population of dogs living now with a hunter-gatherer society in a permanent settlement. What I needed was a remote place with hunter-gatherers living in villages where the isolation would reduce the possibilities that the dogs had been continuously corrupted by any inflow of new genes.

THE MESOLITHIC ISLAND

I found one example of that population of people and dogs on an island off the East African coast. Pemba, located just south of the equator, is an autonomous province of Tanzania. It is 380 square miles (about thirty miles long by over ten miles wide) and sits in the Indian Ocean, about thirty miles offshore. On a clear day you can just barely see the tall buildings of Mombasa to the northwest across the Pemba Strait. The Tanzanian coastline is perceptible in the west. Zanzibar is out of sight to the southwest. Pemba is surrounded by a spectacular coral reef, and could be described as a tropical paradise.

Except that, at the moment, the human population is a quarter of a million people and growing. The people are hunters and gatherers. They eat fish from the sea, and almost anything that moves on the surrounding coral reef. The reef, under severe pressure from so much human activity, is becoming less spectacular by the day. Besides their Mesolithiclike occupations, Pembans also have a foot in the Neolithic world. They raise and eat chickens, cows, goats, vegetables, and other basics, like cassava. They grow some rice and buy a lot from, among other places, India. They cultivate tropical fruits, coconuts, and mangoes. Their once-flourishing spice exports, however, have dwindled to a trickle. Although trade ships stopped at Pemba and Zanzibar for many years, today they are rare. Explorer Vasco da Gama landed here on the last day of Ramadan in 1498, after rounding the Cape of Good Hope, looking for a route to India. Before that, Arab trading routes included Pemba, and many a sultan set up his vacation harem in the region.

In the present world economy, Pemba is out of the loop, an end-of-the-road cul-de-sac. In several areas, children show signs of kwashiorkor, a protein deficiency, suggesting that they live on the margin of what their environment will support. Protein is expensive to import, and thus the coral reef is an important source of protein for them. The big problem is that the reef products are very saleable, and much of them get sold off the island rather than fed to the children.

Pembans, of course, are not exactly a Mesolithic hunting-gathering society. But a great many full-time fisherman travel around the coral reef in dhows—boats made from hollowing out a log and rigging it with a distinctive triangular sail. They fish in the deeper waters at night, using gasoline lanterns to attract the fish, and gather shellfish when low tide exposes the reef. Many details of life have changed since Mesolithic times, but the hunting and gathering out on the reef have existed for a very long time.

Pemba is not exactly a collection of Mesolithic villages, either. Pemban culture might better be described as being on a boundary between hunting-gathering and agricultural—the boundary between the Mesolithic and the Neolithic. We in the United States are almost totally Neolithic, depending entirely on domestic agricultural species for sustenance. We still gather a tiny percent of our food from the wild (berries, ocean fish, caviar), but these items are luxuries and not crucial to our existence. Mesolithic peoples, by definition, were totally dependent on hunting and gathering wild animals and plants. How many people could exist and how well they survived depended entirely on the abundance of wildlife. If a man went hunting and didn't find anything, he went hungry. Pembans today also depend on both domestic and wild species, but if it were not for hunting and gathering, I think they would be in big trouble. And their children, indeed, are perhaps showing signs of those big troubles to come.

Late in the Mesolithic period (fifteen thousand years ago), hunter-gatherer people built stone villages along savannahlike game trails in what is now Israel. These people were called Natufians. They built the first permanent settlements (that we know of, so far) near reasonably constant sources of food. In Namibia, I once saw an example of what these places must have been like. There was an outcrop of rocks in the middle of a grassland. In the rocks was a little gorge with a waterfall and pool. The Namibians hid under the overhangs, waiting for the grass-eating game to come to the only water in the area. Since the game had to come there, they could establish a permanent ambush. While idling away the time, they drew graffiti on the walls. Saving energy by letting the food come to you rather than chasing after it is a good survival trick. It's like sitting in a dump waiting for food to arrive. At my bird feeder is a feral cat that does the same thing.

Similarly, Pembans live on an island with a constant source of wildlife—the products of the coral reef. They are isolated even more, perhaps, than the Natufians. They are surrounded by the ocean. I picture the first villages ever built (like at the Namibian outcrop) as islands in a savannah sea. The permanent village would be surrounded by a sea of grass. Dogs (or wolves) adapting to the dump niche in one of these villages would also be isolated in the vast savannah ocean. Although these oceans could be crossed, it would take energy and skill to do so. Why would a human or an animal want to cross these oceans if it had enough food and water readily available? Why leave the island paradise? The next island would probably have its own endemic population of competitors.

The people of Pemba live in wood, grass, mud, and sometimes concrete houses, but nowadays these also have tin roofs. One reason anthropologists have picked a figure of fifteen thousand years ago for the advent of permanent settlements is simply because they begin to find signs of stone houses shortly after that time. The possibility exists, however, that like Pembans, Mesolithic peoples may have lived in permanent settlements of grass houses for many thousands of years. These villages would be difficult to discover. If that is the case, then dogs could be older than we think. But my hypothesis is that dogs are coincidental with permanent human settlement. It is the change in human behavior that creates the niche that wolves exploit and adapt to. (I'll discuss how old dogs might be in Chapter 10.)

When I first saw the dogs on Pemba, I thought: strays, crossbreeds, mutts. But I soon realized there were no house dogs—no pets—for the Pemba dogs to be crossbreeds of. To an untrained eye, they might look like crossbred mongrels because they are so unremarkable. They all looked just about the same—thirty or so pounds, slender, with short, smooth coats of variegated colors, some with large spots, some with markings on their heads, ears, legs, or tails. Their ears are pendant, or erect but bent over slightly at the tips (tulip ears). As a population, they are amazingly uniform in size. What this all meant to me was an isolated gene pool, not influenced by any other strain of dog that would introduce a variation in appearance. As I drove around Pemba with a small team of students, we were never surprised at what a dog looked like. Nobody ever said, "Wow, look at that big dog!" That in itself was a clue that these were not random strays, and that natural selection was at work. The uniformity of size signaled intense selective pressures, a survival of the fittest for the niche. Any individual much larger or smaller than thirty pounds cannot make it in the niche. There isn't enough food in the dumps to support a big dog, and little dogs cannot defend a large enough home (feeding) range.

In contrast, the *lack* of uniformity in coat color and ear shape signals a lack of selective pressures for those characters. Coat color and ear carriage, those traits that are commonly manipulated "artificially" by breeders, make no difference to individual survival in a natural population.

Pembans, in general, don't like dogs, and it seems to me that many of their reasons could well hold true for the ancients. Pembans are Muslims. Mohammed didn't like dogs, according to Pembans, and Mohammed advised against touching dogs. Evil organisms, they say, live in dogs' noses. That is why dog noses are cold and wet. The nasal discharges and drooling saliva are evidence of disease, insects, and parasites. They believe if they come in contact with dog fluids they can become sick. Back in Mohammed's day and in much of Africa and Asia today, dogs are a significant vector for rabies. In spite of the cute-puppy syndrome, it's important to consider that disease and social beliefs might have been far more consequential in Mesolithic villages.

Pembans believe that God doesn't like dogs. If a dog comes into their house, God will not visit them. If a dog comes into the house even uninvited, the house must be cleaned—*really* cleaned, spiritually cleaned, before God will visit again. They view dogs in the house in the same way I view rats in mine: it is, in my society, a sign that I am unclean, actually or spiritually. If I were clean, these animals would have no reason to visit me.

I was once interviewing a Pemban who liked dogs and said he patted his dogs. I said go ahead, I'd like to see you do it, and he said, "Not in front of my friends."

The Pembans' cultural dislike of dogs probably extends back before biblical times. The Old Testament, shared by Christians, Jews, and Muslims, does not have much good to say about dogs. There are a few cryptic passages that might suggest something nice about dogs, as when Job said, "But now they make sport of me, men . . . whose fathers I would disdain to set with the dogs of my flock." Sheepdogs, maybe? Helpers of people? But most of the several dozen biblical references to dogs refer pejoratively to people who are "dogs"—low, untrustworthy, treacherous, dirty scoundrels. As we might say "You dirty rat," they say "You dirty dog." I always thought it would be appropriate to have a sign at the end of the driveway that said, "Beware of the Dogs (Philippians 3:2)," but Paul's letter was about people who acted like dogs, so I decided against it.

The Old Testament refers to dogs also as scavengers. Isaiah says, "The dogs have a mighty appetite; they never have enough." But here again, he is using an attribute of dogs to describe unpleasant people. When the Bible mentions actual dogs, the imagery is disgusting: ". . . he who dies in the city the dogs shall eat; . . . who dies in the field the birds . . . shall eat." Dogs, then, are city vultures. Most of the descriptions of them are of just that—nasty buzzards prowling the streets at night, ripping apart anything they find lying around, including people. Poor Jezebel. There was nothing left except her skull and the palms of her hands after the dogs got through with her.

The three major Middle Eastern religions all have taboos about dogs. Their food laws are explicit about not eating dogs, although other cultures find them a delicacy. Some North Koreans building a football stadium at Chake Chake (on Pemba) actually ate some of the local dogs, much to the mirthful revulsion of the people. Being reminded of that incident fifteen years later, Pembans still made the same face that Americans make when you mention eating rats: "That's disgusting!"

Those cultural taboos well could have started with the first Mesolithic villagers. Certainly, the ancient attitude toward wildlife in general couldn't have been too friendly. In their book on African wildlife, *Wildlife, Wild Death,* Yeager and Miller describe the typical East African attitude toward all wildlife as pests, which you occasionally eat, and which occasionally eat you. Only recently in our culture has the wolf been anything other than big and bad. Wolves, coyotes, and jackals have traditionally been seen as vermin, purveyors of disease, and predators of domestic animals. The concept of the noble wolf is for the most part recent, and not well accepted worldwide.

With Pembans holding such attitudes about dogs, you would think there wouldn't be any dogs on the island. (If Americans have the same attitudes about rats, you would think we wouldn't have any rats.) In fact, when I first went to Pemba, I was told not to expect to see any dogs. I offered a quarter to anyone in our group who could spot a dog. I nearly went broke in the first two hours. Pemba was loaded with dogs, so many that they are often a problem for the people. In fact, when the dog population gets too high or troublesome, or there is a rabies epidemic, the army is called out to shoot all the dogs. But they never get them all. It is like trying to eradicate all our rats; it is practically impossible.

At first, while wandering around Pemba, I thought the dogs appeared to be pets. They behaved like pets. I saw dogs sleeping in the village square, or in front yards and around homes. Were they pets? Occasionally we tried to get close to a dog—close enough to pat one. This was hard. They always moved just a little bit out of range. We would approach very slowly, cautiously, and then the dog would shift just enough. Once, on the island of St. Kitts in the West Indies, I reached out slowly to stroke a dog behind the ears. Skinned-back lips appeared, and I understood that fondling was not part of our relationship. For the most part, the Pemban dogs paid little attention to me, except to move slightly away—just enough. If I pressed my attentions, they trotted off. Pigeons in the park behave the same way. Trying to pat a pigeon is like trying to pat a Pemba dog. Calling a pigeon in the park is like calling a Pemba dog.

When I talked to the Pembans about the dogs, I had to be very careful phrasing my questions.

"Is that your dog?" I would ask. "Yes" would be the answer. "Do you feed your dog?" "Yes." But "What do you feed the dog?" got a look of confusion. "Could you call the dog over here?" again got me the look! "Of course I could call the dog over here." "Will he come?" "I don't know, I never tried." "Well, call the dog." "What should I call him?" "Does he have a name?" "I don't know," followed by giggles. "Well, call him over." My informant would say something in Swahili. Nothing happened. "Can you go over there and pat the dog?" That got a big giggle— why would anybody want to do that? "Have you ever touched him?" "No!"

Why would a Pemban say this was her dog? Because the dog was always in or around her yard. "Is that your tree [in the front yard]?" "Yes." "Do you water the tree?" "We dump the dishwater there." "Could you call the tree?" "Sure—what do you want me to call it?" Dog ownership was like yard tree ownership. Residents had nothing to do with the tree being there, but it was in their yard. However, chicken ownership was emphatically, "That's mine." Chickens have value. You can eat chickens and you can sell chickens. On an island where protein is in short supply, chickens are a prize. On an island where food is in short supply, people cannot afford to keep a pet dog, unless it doesn't cost anything, and doesn't compete for food. In other words, the dog has a commensal relationship with the people.

The dogs on Pemba are not fed, but they do feed. They scavenge the wastes from the villagers' activities. Some dogs hang around dwellings, but some gather around groups of people. When the fishing boats land in the early morning to auction off the day's catch, dogs are there.

Groups of fishermen collect in the morning sun on the beach by some lonely tree and tell the night's adventures of suckering a billfish to the boat with the light from a gasoline lantern. Listening and watching will be a dog, sometimes two, apparently asleep in the shade. As the fish are gutted out and divided into parts, the dogs are the recipients of the waste.

The men brew tea and eat bread. Sometimes, someone will actually throw something to a dog. It seems almost like throwing a coin to a beggar or throwing crumbs to a pigeon. It isn't exactly like feeding the dog or even having any empathy for it. Feeding pigeons in the park isn't exactly about feeding pigeons, either, but more like watching pigeons eat—because they are there.

Around the houses, dogs and chickens behave exactly alike. They camp in the yards. They eat discarded scraps. The waste scraps the chickens eat are much smaller than the dog scraps. Chickens find lost millet or rice seeds, and little insects. They wander along the road and eat tiny things that have been squashed by passing vehicles—insects and seeds, which are below the size limits of what the dogs feed on. Chickens are yard scavengers in the same way the dogs are. But they live in different niches—chickens and dogs do not compete with each other for the same village wastes. Perhaps chickens domesticated themselves in the same way dogs did. I should add chickens to the list of invaders of permanent settlement, along with rats, pigeons, and cockroaches.

The dogs on Pemba tend not to eat living things. They behave more like buzzards. They don't waste energy chasing things. They spend time hunting (searching) and gathering. It is not a constant motion. In fact, there is not much motion at all. Finding food means positioning yourself in a place where food will show up. Vultures position themselves over a dying animal, waiting for food to "show up," so to speak. The dog positions itself at the refuse pile waiting for something dead to show up.

Each house in a Pemba village has its own dump and its own latrine; both are located in the backyard, which, like many backyards, abuts somebody else's backyard. Dumps on yard boundaries often merge together. One or two dogs might command several of these dumps and latrines. A small boy guiding me into a latrine in Turkey one night turned abruptly and, running smack into me, yelled in Turkish (which I hadn't understood until that very moment), "Run! The dogs are in the latrine!" Turkish dogs, feeding in a latrine, might be defending food, but it was more likely that we just trapped them as one would trap a pigeon, by coming through the door. Whichever, it is a great time to get bitten. Pandemonium broke out as the two of us ran up an alley, apparently chased by the two dogs, but they turned in the opposite direction from us when we got out to the street. They were executing the same escape behavior we were, but circumstances had put them behind us in the narrow alley.

It is well known among locals that many dogs in this world feed at latrines. I need not say too much more about the coprophagous dietary habits of dogs, except to note that, in some cultures, there is a mutual benefit for dogs and people in this respect. For instance, while in Kenya working near Turkana villages, our daughter, Karyn, noted a practice that anthropologists have been reporting for decades. At birth a baby is "given" a puppy as a substitute for baby wipes.

Not much "wolf" behavior still exists in village dogs. Pemba dogs seem to have territories for feeding activities. They don't exactly threaten other dogs, nor do they need to. These are very low-key animals. They simply space themselves out around the environment. The same dogs will be in the same place day after day. They may search for food over a wider area at night, but we didn't see it. (Crawling around people's backyards at night has its own hazards: "Hey, what is going on out there?" "Hi, we are watching the dogs eat." "Ah . . . ah . . .")

Neither is it obvious to me whether the daytime territory was part of the feeding territory or part of a separate reproductive territory. With many canids it is hard to separate the two. Wolves, for example, have a feeding territory, where they also den, and where they have a rendezvous point for feeding their pups. Howling is thought to mark these feeding territories, but howling increases during courtship periods. Coyotes, too, yelp more at courtship time. My kennel dogs bark and howl all the time but it gets a lot worse if there is a female in season.

Unlike wolves, Pemba dogs bark, especially at night. For them barking seems to be marking a location: "I am on site." "I'm here, this is my spot." No reply is necessary or expected. I'm not aware of any formal studies about the possibility that barking in dogs is analogous to howling in wolves as population assessment behavior. There are such studies for other species that find that the noise level at roosting areas has an effect on breeding success. But I absolutely believe, when my yard dog barks at night and I can hear the dog at the next farm also bark, that the next dog's bark is answered by one farther up the line, and so on until the bark travels all around the world. That is why, after relaxing into the quiet of the night, my rest is again disturbed by my dog barking. The bark has come all the way around, as has the night.

In New Guinea, the local "singing dogs" maintain their territories outside the villages. They "sing" a special song to mark these territories, with trills and patterns similar to birdsong. Even in our kennel at Hampshire College, the singing dogs produce their musical repertoire, unusual in the dog world: modulated yodels and, best of all, a high-pitched trill. Because their songs are reminders of birdsong, they suggest that their territories are reproductive, not feeding territories. Maybe this is another clue that the static, continual nighttime barking of Pemba dogs has a reproductive function. The New Guinea dogs come into the villages and feed on the wastes that fall from the stilt houses above, and as I understand it they are quiet while feeding.

Often it is thought that barking is the alarm of a watchdog, warning the master of prowlers. On Pemba there are no masters, and still the dogs bark. But barking does warn people about prowlers. Barking increases in intensity if an actual prowler is detected. People can easily distinguish the difference between just barking and barking *at something*. Experience teaches people that when the pitch and rapidity of the bark rise, something or someone strange is present. The difference between *ruffruffruff* and *R O U F R O U F R O U F* is significant.

This does not mean that the function of barking is to warn people. Nor does it mean people selected dogs to bark at strangers. Many species recognize and appropriately respond to the alarm calls of other species. In the forest, people will respond to the alarm calls of blue jays and crows, and yet nobody believes people selected birds to direct alarm calls toward strangers.

Most of the Pemba dogs live alone or in very small groups—at the most, three. If there are three dogs together, they are often the same color, perhaps indicating a mother and offspring or some other genetic relationship. Notice that the word is "group," not "pack." Pemba dog group behavior does not mimic wild wolf pack behavior.

In fact, contrary to popular belief, dogs around the world do not (or only rarely) exhibit "pack" behavior. Wolves pack cooperatively during hunting in order to kill large prey. Upon killing, pack members return to the den to regurgitate food for a single litter of puppies. Usually, no single wolf or pair of wolves can accomplish either task of killing large prey or raising a litter of puppies alone. Thus the pack—often composed of family members—works together as a survival strategy.

Dogs hardly need a social organization to feed on discarded chicken bones and mango skins. For dogs, other dogs are no help when it comes to feeding themselves or feeding pups. In fact, other dogs are not only no help in finding garbage, but they are the chief competitors for a limited quantity of food. Thus, packing behaviors are not to a village dog's selective advantage. There are few benefits in getting together to feed, and no motivation to feed someone else's pups.

How important this observation is for our understanding of the dog. The village dog is not a pack animal in the same sense a wolf is. Perhaps packing behavior has even been selected against in the village dog. Although dogs are individually territorial, inhabiting a solitary space, they are obviously not asocial. They evolved their own adaptations to feeding in the village—small head, teeth, and brains—and their feeding behavior is specialized to their scavenger niche. That means they search or wait for food alone, not cooperatively. And they are "aware" that humans are the source of that food; thus they focus on human activity rather than trying to avoid human activity.

Is packing behavior genetic? Research indicates that packing behavior is a developmental response to a specific habitat. Wolves don't always pack; some populations never pack. Coyotes, which aren't thought of as a packing species, often do pack, especially in places where they are not disturbed by wolves or people. It is doubtful that Pemba has any of the environmental signals that elicit packing behavior.

I don't see much in dogs that indicates they have the fundamental behaviors that would allow true wolflike packing. For example, unlike wolves, male dogs do not as a rule take care of or regurgitate food to pups. Other behaviors of dogs indicate that they are not disposed to the kind of social organization that adult wolves have. In fact, it is doubtful that female village dogs regurgitate consistently enough for pup survival. Parental behavior in dogs relies on the dump being there for pups to forage in. Dogs are adapted to a very different niche than wolves, and their social behavior has likewise evolved so that it is appropriate to that niche.

A modest exception, and an illustration of how environment can alter foraging behavior, can be seen in the dogs at the Chake Chake city dump on Pemba. Here is a disagreeable bunch of dogs. At Chake Chake the dump is very close to the abattoirs. Something quite different goes on here. There is much more food—scraps of meat, bones, guts, heads with contents—in a much smaller area. Besides dogs, there are hooded crows, dozens of little birds, and a host of other characters scrambling for rich morsels.

There is an important doggy axiom here. Rule: As the quality of food increases, aggression between dogs escalates. Each individual dog will take bigger risks to get quality food, and they will fight over it. In our kennel we could control aggression by feeding the dogs pig pellets instead of high-grade dog food. Pig pellets are not bad nutrition, but our dogs just didn't like them. This allowed us to group-house some livestock-guarding dogs, with no fighting at feeding time. But throw a steak in there and, *shazam!* an instant and ferocious fight.

In general, the village dogs of Pemba are in fairly good shape. A few are thin. There are pregnant females and puppies. The claim that an ever-increasing population needs to be thinned out from time to time suggests that reproduction is successful, and that as a population these dogs are viable.

The dogs of Pemba could be described as naturally tame, and fit very nicely into Belyaev's observations about selection for shorter flight distance. The people of Pemba provide, unknowingly, a niche. Into that niche move dogs, and within it they find food, reproduce, and avoid hazards, just like any wild species. They are still wild—as wild as pigeons or rats. They live symbiotically with people, they depend on people (to provide the niche), and they benefit from people. The people derive little benefit from them. It is a commensal relationship. If the people ate them regularly, as they do in other parts of the world, then it might be considered a mutual relationship. Chickens on Pemba have a mutual relationship with people.

Pemba is perhaps an island paradise for the commensal dogs. There are no chains and collars or fenced yards or mandatory walks on a leash. They aren't spayed or castrated, locked in homes, or isolated from their social group; nor do they need to be housebroken or attend obedience classes. They don't have any human trying to be leader of the pack, or pretending to be Skinner with punishments and rewards, or Pavlov with click-and-treat training. These dogs are free of all that. They are free to live the natural (wild) life of a basic dog. It is no more or less meager a life than living wild and competing for big game—and not nearly so dangerous.

Are not the Pemba dogs—and village dogs the world over—the original dogs? The criteria set forth at the start of this chapter about canid traits and environmental events hold true. The commensal relationship we can see today supports the concept that an early protodog could well have survived and even thrived in Mesolithic villages. These dogs are not simply village strays. I believe they are descendants of the first-evolved domestic dogs from the Mesolithic period of human history.

I also believe that similar populations of early village dogs are the roots of our modern breeds.

A marketplace in Canton, China. Delicacies such as rabbits, chickens, turtles, or dogs are for sale in this Cantonese market. Among dogs, puppies command a premium price because they are so delicious. In some areas of the world the dog is just another domestic animal, valued as food and skins for quality garments. (Photo by Gary Hirshberg)

CHAPTER 3

Natural Breeds

HOW DO DOGS GO from the locally adapted village scavenger to the hundreds of breeds we now recognize? Most of us have assumed that breeds are the result of intentional human selection, what Darwin called artificial selection. But that is not necessarily so. Natural breeds can arise locally with no human interaction. Or people can inadvertently create a breed without intending to. It is also true that people can consciously create a breed, as we will see in Part II.

Doesn't the term "breed" necessarily imply a mutual relationship between people and dogs? Mutualism occurs when both species benefit from the relationship. Many of us assume that if there is a breed of dog, then it was selected to benefit humans in some way. But again, these are assumptions to be tested, not forgone conclusions. It is possible to get breeds by natural processes.

I'll explore natural processes first. Do the Pemba dogs qualify as a pure breed? Well, why not? Don't they fit the requirements of a pure breed of dog? Their appearance and behavior are as uniform as any breed's. They are sexually isolated from other dogs because they live on an island. All they need to be a "recognized" breed is a breed club with a published breed standard and someone to keep genealogical records for seven generations. In many cases, this is exactly how breeds are founded.

I can see it in the breed book now.

THE PEMBA HOUND

Origin. This breed dates from the very earliest biblical times; their images have been carved on pyramids. They are mentioned in the Bible. It is said they arrived on Pemba in hollowed-out logs called dhows, to help the sailors with lunch. Several specimens have been found in Stone Age Natufian graves. These dogs, lying in affectionate repose with their loving masters, have retained their companionable qualities down to the present. It is thought by at least one scientist (me) that this breed is the direct descendant of Middle Eastern wolves.

Description. The Pemba dog is a medium-sized hound with a slightly elongated body. The breed standard calls for males to be 16.5 inches at the shoulder; females are slightly smaller. They should weigh between 22 and 30 pounds. The head should be well proportioned, not too large, with round eyes and a sweet expression. Pricked ears are preferred but other ear carriages are acceptable. Coat color should be variable; unusual patterns are highly prized.

Personality. This hound is a tenacious hunter of food, very intelligent but somewhat standoffish. Around the yard it is gentle and agreeable, always present and extremely loyal. At rest, it is serene and happy. Trustworthy with children and chickens.

Uses. The Pemba hound is most useful in keeping other dogs out of the yard. It is ever-vigilant and performs its job in any kind of weather. It is a good watchdog and barks at strangers, especially at night, when strangers are most dangerous.

What more could anyone possibly want? Some enterprising soul adopts a few specimens from a local population of free-living dogs, attracted by some quality that might be useful, or interesting, for whatever reason. Perhaps the selected dogs are distinctive in some way: they are barkless, or vocalize with trills, or are white. The barkless basenji and the New Guinea singing dog are relatively new arrivals to the West, brought in as natural breeds, as unusual "subspecies." In the American South, little pariah dogs called the Carolina dog are being adopted from the wild as pets, and have started along the road to becoming a legitimate breed.

Thumbing through the dog books, one finds many breeds whose descriptions parallel the one I just wrote for the Pemba hound. Several of the Swiss hounds (whose descriptions strongly influenced what I wrote) were from the Nile River in Egypt, and they look very village-doggy to me. The Canaan dog description also could be used, except they are slightly larger. These are not isolated examples. Several dozen breeds greatly resemble in many details the Pemba dog. Someone found what he thought was an original breed of dog in some remote village, declared it was a breed, and took the next steps to turn it into a kennel-club dog.

The question is, why does the traveler think this remote village population of dogs is a breed? I think it has something to do with the uniformity of the dogs in the population. Just as I noted that the Pemba dogs were uniform in size and shape, travelers can find a remote place where the dogs are uniform in other characteristics such as color. They find a village and almost every single dog is white. They therefore assume that people must have bred these dogs to be white, and purpose-bred dogs begin to sound like a breed.

However, natural processes can produce, could produce, and do produce populations of unusual and uniform dogs, that is, dogs with a distinctive conformation. To me, these are the natural breeds.

We have already seen examples where natural selection has shaped the heads, tooth size, and brain size of village dogs. Each population of dogs is selected to be hardy and efficient in their particular niche. All around the world, dog niches vary in quantity and quality of food, nesting sites, or hazards, and local populations of dogs have adapted to these conditions. In the North, village dogs tend to be bigger, adapting them to colder climates. The different selective pressures will shape dogs in different ways.

There are other natural mechanisms that create breeds of dogs. Catastrophes, such as diseases, periodically wipe out entire populations of village dogs. The dogs on Pemba and in much of East Africa are frequently ravaged by rabies. Many dogs die of this disease, and many more get killed by people who try to eradicate the dogs for their own protection.

When a population on an island such as Pemba gets entirely wiped out, eventually some Adam and Eve dog pair migrate in and start a new population. Or the population can be wiped out except for a single pair (or pregnant female) that start a new population of dogs, all descended from the original pair. These may be exaggerated examples of what does happen. But whatever the number that survives or migrates to the canine wasteland, it leads to an interesting phenomenon.

Our founding pair do not, indeed cannot, represent the entire genetic variation of the population they came from. Biologists call this circumstance the "founder principle." The new population of dogs will be more similar to the founding animals but cannot resemble the whole population that has died out. If the founding animals just happened to be white, then their descendants could all be white, although the parents' population may have had many colors and patterns. In biology, the predominance of white in the new population is called a founder effect.

If, for example, I reach blindfolded into Madison Square Garden in New York City during a basketball game and select two people, one male and one female, at random, it could be that just by chance I happen to pick two with blue eyes. If I release these two people on an island where there are no other people, in one hundred years the resulting population would be predominantly blue-eyed, in spite of the fact that the population at the basketball game had 50 percent brown eyes. On the island, the frequency of the blue eye color is not the same as in the population the founding pair came from. Indeed, we might say that this island has a race of blue-eyed people. If it were dogs, I might say I had a breed of blue-eyed dogs. However, it would be a mistake to assume immediately that people had bred for blue-eyed dogs, and also a mistake to assume that the blue eyes were an adaptive advantage.

Regionally, breeds of dogs with distinctive traits can be created and re-created by natural selection occurring in response to changing conditions and to founder effects of periodic population catastrophes. Periodic epidemics are the rule in animal populations, not the exception. Populations expand, overeat their food resource, and succumb to some pestilence. Wolf, coyote, and fox populations periodically get mange, and starve or freeze to death, or die from secondary infections. Rabies, distemper, and parvovirus are common ailments of canids, and have disastrous effects on a population. Pups in any given year can be weakened by a bad season and poor prey availability, and thus vulnerable to hazards.

For a population ecologist, these events are normal. Each time a pestilence severely reduces a population, I would expect to see a founder effect in the future population. I would not expect that the gene frequency for any character would stay the same for any long period of time. Therefore, I would not expect that any natural breed of dog is very old. Certainly I would never expect to find an ancient breed of dog, in the sense that the present population represents the gene frequency of its ancient ancestors. My Adam and Eve dogs may be the direct descendants of an ancient breed of dogs, but they cannot be representative of an ancestral population. Neither are their offspring.

When one understands that there are periodic shifts in gene frequency, then one realizes why it is so difficult to identify specific genes for pit bulls, golden retrievers, or any other breed of dog.

No breed can be ancient in the sense of an unchanging gene sequence. Even within our modern breeds, the gene frequencies are constantly changing, sometimes due to natural selection, sometimes due to artificial selection, and sometimes just due to chance events. The breeders themselves shift the gene frequencies in many ways—breeding many females to this year's grand champion creates a founder effect in the next generations, for example.

PEOPLE BECOME CONSCIOUS OF DOGS

Up to this point in the story I have tried to keep people out of the process of dog evolution and breed formation. Now is the time to add people to the equation.

Turning basic village dogs into pets is not only easy, but I imagine in many places it is practically impossible not to. Once the basic commensal dog exists, the transition into functional breeds with mutual benefits for both dogs and people is easy. The following chapters on livestock-guarding dogs, herding dogs, and sled dogs give the details on how highly functional mutualistic breeds are created from village dogs.

But first, now that we have transformed the wolf into the village dog, we need to look at other mechanisms for transforming village dogs into breeds. We can now suppose the adoption by people of cute puppies—but these are not wolf puppies; they are the offspring of the naturally tame, the *genetically* tame village dogs born in or around the village. These are the pups that get adopted into households and cared for by people.

My favorite account of adoption took place on Zanzibar. This win-win story has me sitting on a beach chair at a swanky seaside resort waiting for the Canadian documentary film team to fly in and photograph village dogs. On the beach are three dogs, and having nothing better to do, the inveterate ethologist starts the follow-them-around routine. How do these dogs earn their living? I muse.

The dogs can be touched—but ear scratching is about the limit. They obviously "belong" to the hotel. (As it turns out, the hotel belongs to them.) They spend their days on the beach, often under the beach

A Mexican boy and his puppy. Adoption of wee puppies, often by children, leads the commensal village dog toward a mutualistic relationship with people.

chairs for shade, even when people are sitting in them. They rarely come off the beach into the grass-roofed cabana during the day, and hotel guests lunching on the beach share tidbits with them. They move into the hotel yard at night, right after the bar closes.

Every night, the routine plays like clockwork. All evening people sip drinks, laugh and talk, maybe dance a little. You can see the gasoline lights of the hunter-gatherer fishermen at sea. It is an adult group, sometimes parents with small children, and so by 11 P.M. the place is pretty quiet. The staff finishes cleaning the dishes from dinner, and shuts all the doors. They go home. The very last event is the now solitary barkeep locking the liquor closet.

Two milliseconds after he turns that key, with its big metallic *click,* all hell breaks loose. It is the original "click and treat." Pavlov would have had a field day here. Thirty-odd cats come alive; they emerge from everywhere. Fighting, wailing, spitting cats. The three beach dogs materialize at the kitchen door, where a wheelbarrow full of scraps has been put out to wait until morning to be dumped. All night and into the early morning a variety of animals works the hotel. In the rafters of the open cabana are a couple of tiny monkeys that have snatched a share. In the early morning the hooded crows, little gulls, and English sparrows all have a special place to work—their own hotel space.

At sunup it is a Mesolithic world. The only people visible are in the tiny dhows, coming in with the night's catch. The crows have finished feeding and are pulling the stuffing out of a beach pillow for nesting material. The gulls and crows rarely come under the grass roof, for that is the domain of the English sparrows, monkeys, and cats.

The three dogs are up early and are very active—as active as they ever get. They groom and play a bit. And usually they hang out at the tide line. Why the tide line? Because next door, to the east, at the sultan's abandoned palace, three other sulky dogs watch from the end of the concrete wall dividing the two properties. West along the beach are two other dogs that are often on the boundary line. They occasionally burst across and run down the beach in front of the hotel. But the hotel trio rushes with them and "plays" with them. It isn't a game, however—it is a test; they are really playing chicken. The event is a territorial conflict. But it occurs daily, even though nothing much ever happens, except that the ritual is performed. And that is it for the day. The three hotel dogs spend the rest of the day sleeping on the beach with one eye open, as do all the neighboring dogs.

These hotel dogs have a really good life. They have both quality and quantity of food. All they can eat. The food is in a single wheelbarrow, which appears at a precise time and is easy to defend. Whatever is left disappears later in the wee morning hours when the cleanup staff remove it. Nothing is left to defend, making life for the dogs that much easier.

I ask our proprietor, "Tonino, are those your dogs?" "Yes!" he says with affection and pride. "Do these dogs have names?" At which he grins broadly and says, "Of course they have names!" Scarface, Spotty Dog, Mommy Dog. Tonino tells me the story of how he acquired his dogs.

Less than ten years ago, rabies broke out on Zanzibar. During the epidemic, as usual, the government shot "all" the dogs. Shortly afterward, the current trio of dogs showed up at this beach resort. The European hotel management could see the handwriting on the wall. Not only was rabies a recurring theme, but a resident population of flea-bitten, rib-showing, mouth-frothing, fly-snapping dogs was not great advertise-ment for a top-end resort. At the very least, European guests would think these people didn't take care of "their" dogs. The problem was, if you just got rid of them, a new group would move right in. Besides, the hotel managers liked dogs. They thought that because they were black and tan, they must have some Doberman pinscher in them, maybe crossed with some Italian greyhound.

Obviously, enlightened dog management was needed. They decided to vaccinate all "their" dogs against rabies. It wasn't easy. They hired a vet-erinarian, who had to use tranquilizers and then follow the dogs until they could be captured and handled. They vaccinated against rabies, and while they were at it, they thought, why not distemper, hepatitis, and lep-tospirosis. Why not worm them? And then the brainstorm: Why not spay the females? A temporary operating table was set up and the females lost their reproductive potential.

It turned out to be an excellent strategy, based on sound theory; keep the resident population alive and healthy so they can defend the hotel from the cur dogs next door. No matter where you are in the tropics, a dog will show up whether you want one or not. It is better to take care of the one you have so you don't have to have more than one. That's the small beginning of a mutual relationship.

Now, Tonino had another problem. He had European guests and they didn't understand that these were wild animals. They thought they were dogs, and so must belong to someone. The problem was, how to make them attractive to the guests. The dogs had to be trained to behave like pets. It took a differing amount of time to tame/train each one. Scarface held out for a year—the longest. Tonino and his col-leagues conditioned the dogs to allow an occasional pat, to exhibit a calm, refined begging behavior, and to accept handouts politely.

One good feeding a day means the dogs don't have to search contin-uously. They can sleep conspicuously on the beach among the guests and defend their paradise. And there is no increase in population in response to the overabundant resource. Yearly, the vet comes with the booster shots and worming pills, and the population at the hotel remains stable. He still needs to tranquilize the dogs to administer the treatments, for like any wild animals, they are shy about being captured. But the point about the dogs of Zanzibar is that they are dogs, and are tameable and trainable. They are wild, but they are not wolves.

Tonino's story points out that the benefit for him of adopting dogs is self-protection. He takes care of the dogs so the dog problem doesn't get worse for the hotel. In the beginning, the hotel didn't want dogs. In the end, Tonino shows a pride in "his" dogs. And in the middle, the hotel dogs have fared better than their colleagues. They get board, room, and health care—all the while using appropriate dog behaviors that mesh with the needs of the people.

Once I was in a small Turkish town looking for dogs to buy. The locals told me where I could find a litter. As I walked down the street, a woman came out of a house, grabbed a puppy from out front, and hurried away with it. A favorite puppy, obviously, that she didn't want sold. She liked that one for some reason—it was cute, feisty, or some unusual color—take your choice. It didn't matter why. In this village the dogs were still living primarily on waste products. It looked like several dogs had distemper. But, for whatever reason, she chose this pup, thereby conferring on it a better chance of survival than its littermates. In giving this dog a little extra care, she tipped the balance toward its survival.

At the same time, whatever trait distinguished that pup from its siblings stands to survive into the next generation. Here is a nascent evolution of a breed quality, perhaps. Let's say she picked the dog because of its color. Now that dog has a better chance than its siblings of passing its color to the next generation. Thus, the woman wasn't exactly breeding dogs, but rather creating a differential mortality by supporting one color over another. I suspected, because of the tremendous color variation, that there was no selective advantage of one color over another. The woman, by focusing on color, was not interfering with truly adaptive traits, but with a superficial trait. This is an important point to remember, particularly when in a later chapter about today's breeds I discuss what can happen when people interfere with adaptive traits.

There are countless variations on scenes like these. Tonino's dogs selected him (or his environment), while the Turkish woman picked and chose the specific pup intentionally. Biologist Alan Beck noted that people in Baltimore fed stray dogs differentially, directing their offerings toward a favorite dog. This behavior is something like throwing a peanut to the white pigeon.

If the village pup feeders are sorting through the variation in pups, then we have artificial selection occurring even though nobody is breeding dogs. For artificial selection to occur, whatever variation is being selected must be observable in the pups. A pretty coat is an obvious and distinctive feature. Another is a cute soliciting behavior by the pup that comes and explores your hand, and so gets picked over one that cowers in back or snarls at you.

Two things are happening here with these adoption processes. First, the dogs are evolving more tameable-trainable personalities, and people are facilitating that evolution without ever purposely breeding a single animal. And second, whatever distinguishes the chosen puppy is more likely to appear in the next generation, and so that characteristic increases in the population.

Locally, the dogs take on a distinctive form, color, or behavior. Maybe, as in Tonino's case, people notice that their resident dog or favorite dog is useful. The Masai in Africa have a little dog about the size of the typical village dog. The children play with these dogs when they are puppies. When young boys take the cattle to graze and water, the dogs go with them. If a lion shows up, the dogs bark at the lion and the boys are warned. I talked to Masai warriors who have a favorite story about a boyhood time when their dog led them to a cow-killing lion. The conclusion of the story is (always) that the boy kills the lion with his spear. The boy becomes a man, in part because of his relationship with the dog.

No one breeds these dogs, in the sense of arranged marriages, nor do they feed them the way we think of feeding a dog. They are village dogs that get played with as pups by children (tameable) and follow the drovers (trainable), and then when a lion shows up, they behave like dogs (fearful barking), which the people learn to respond to. Hence the dogs are useful.

The Masai dogs tend toward a reddish-brown color. Founder effect? Maybe, but I think the Masai prefer the reddish color because it matches their red clothing. I bet they give a little extra support to individuals with this color. They call them Masai dogs; we call them Masai cattle dogs. I'd call the Masai dog a breed or race. It is a nonrandom distribution of red alleles, geographically based. People have become one of the agents of natural selection by caring for the red ones.

To me, the Pemba dogs and the Masai dogs are examples of perfectly good breeds. Breed genesis starts with the isolation of a population, and

A Masai boy and his working dog. These dogs actually get attached to the cattle and accompany them even when the herd gets a new young shepherd.

that can happen naturally (with no specific intention by people) or artificially (with specific intention). The isolation of populations can be reinforced and speeded up through adoption of preferred animals. Human support augments the "natural" differential mortality. People may, at first, select traits in these dogs for capricious reasons—they might select for reddish dogs, or for dogs always tagging along on a walk or hunt.

Modern breeds have had similar histories. Just one example of this highly typical method of creating a modern breed is the golden retriever. According to the 1966 handbook of the Golden Retriever Club of Scotland, the golden is generally accepted as "descended from rare yellow 'sports' born of black wavy-coated retrievers." The first yellow retriever to figure in the development of the breed was a single yellow pup born in a litter of black, wavy-coated retrievers. That male, Nous, was mated in 1868 with a Tweed water spaniel, Belle. The resulting litter contained four yellow retrievers, which became the root foundation of the breed. To further his planned yellow-line breeding, Lord Tweedmouth, an avid sport hunter and careful dog breeder, kept detailed records about the evolution of the golden dog. Further outcrosses in that generation with red setters, black retrievers, and Labrador retrievers in the mix led eventually to those goldens whose pedigrees, starting in 1901, are the oldest in the purebred stud book of the breed.

I have no question about the intentionality of the people here or their abilities to sexually isolate and arrange breedings. But I would like to ask a couple of questions. In this case there was an animal with superior behavior that by chance also had a unique color. It is also possible that the unique color of Nous conferred favoritism on him, which in turn enriched the development of his superior qualities. Was it nature or nurture that led to his superiority? Lord Tweedmouth offered no conjecture. But Nous was not only supported, he became a favored stud. Nous was being bred to superior animals to produce superior puppies.

There are two assumptions here: 1) the superior ability is genetic and 2) the pups inherit their superior behavior from Nous. However much fun and romance is involved in these discussions, I'd like to offer a counterhypothesis: Nous was a favorite simply because of his color and got more attention. He was bred to good dogs of other breeds and the pups were superior because of hybrid vigor, because their mothers were superior, and because people expected them to be superior and therefore favored them.

I think coat color has a huge selection value for who gets bred to whom. Humans are capricious and have good abilities for seeing color. For dogs, who have either terrible or no sense of color, coat color doesn't mean anything. The sole advantage of the unusual coat color is that it gains the individual dog support from humans. In the natural breeds, human fancy for a distinctive color is of little harm because the dog still has to make it on its own, as do its offspring. Favored dogs of the natural breeds still breed on their own, and survival of the fittest will still prevail. But when people finally get to artificially isolating these animals and sorting for capricious reasons—then dogs are in trouble. Arranging matings for capricious reasons, I think, is a major mistake, which we deal with in Chapter 7.

But that is anticipating a later chapter. What is important now is to understand the basic dog and a few possible ways that it evolved into the vast varieties of breeds. The next chapters, about how those natural breeds became great working dogs, describe the dog's progress from commensal village scavenger to mutually beneficial symbiont.

PART II

WORKING DOGS AND PEOPLE: MUTUALISM

In Part I, we looked at the evolution of dogs from wolves. We saw populations of dogs living near humans, scavenging on human waste, but without any direct human assistance—and thriving. We could even account for different races and natural breeds of dogs evolving without human interference. These village dogs have a commensal relationship with humans. They gain benefit from humans but they don't cost the humans anything. Their foraging, reproductive, and hazard-avoidance behaviors are similar to those of wild species, adapting them to survive in a niche. The niche in this case was inadvertently created by humans who had begun to live in permanent settlements, which allowed for the accumulation of waste food.

The dog is simply a derivative of the ancestral wolf, a new form that is adapted to feeding on a new source of food. The size and shape of the dog are a response to this new food source. The dog's behavior, its orientation, and its movements in concert with humans are also an adaptation to this new food source. The other, inexplicable differences between dogs and wolves can be viewed as genetic leaps—developmental accidents—precipitated by the rapid evolution for the new, tamer feeding behavior.

The concept of "village dogs" gives an unusual picture of dogs. Dogs are usually thought to have been created by humans to work or hunt for them, resulting eventually in specialized breeds. Village dogs are seen as dogs that have strayed away from their lives as purebred pets and breed with other strays to produce mongrel offspring, making their living by upsetting trash barrels. Yet it seems likely to me that these millions of ubiquitous strays are not discarded crossbreeds, but are actually the original nonwolf, the village-adapted form. From this population people capture individuals for pets, sometimes training them to do tricks or perform little tasks, and then breed favored animals together to develop useful breeds. It is from this basic population that our breeds, known for their extraordinary and highly specialized abilities to serve people, have evolved. I once saw a trick dog performance at the San Diego zoo. When the trainer was asked what kind of dog it was, he replied, "A Canardly—I can hardly tell what breeds are in it."

With the evolution of the working breeds, the status of the dog changes from the commensal scavenger to the symbiotic relationship known as mutualism. Whether it is a true mutualism, where both species become truly dependent and physically adapted to the relationship, is not my question in Part II. It is probably not ultimately to the dog's biological benefit to enter into this relationship, but I will discuss that later (Part III). With the creation of breeds, the original adaptive form of the dog is changed into something that now pleases humans. It might be argued that the dog becomes more dependent on humans and thus vulnerable to human capriciousness. At the same time there is no related change in the form of humans. Therefore the association is not true mutualism in the biological sense.

My job in Part II is to illustrate the biological principles that drive the formation of differentiated working breeds. Why, in other words, do breeds look—and behave—the way they do? The tasks they perform are sometimes difficult to imagine—a carnivore canine that protects the prey species from another canine? Here we have sheepdogs, specialized carnivores, actually herding sheep or protecting them from other carnivores, and not eating them themselves. The wolf ancestor must be turning over in his grave at the sight. What is the reward for a dog to herd or guard sheep? Other specialized dogs will pull a sled loaded with supplies. Sled dogs work very hard, and for what? Certainly they are not running because someone is dangling a steak in front of them. They run, pulling sled and driver, until the driver says, "Whoa!" Retrievers swim out in cold water and fetch a very edible bird and return it to their handler on shore, intact. Herding dogs chase sheep but do not catch them. All they get is a quick pat and a quiet "That'll do." Do they know they'll be fed that night? Even if they do, performance is unlikely to happen for such delayed gratification.

For many of the tasks the different breeds do, often working to exhaustion, there is no apparent immediate reward. That is very unusual behavior in the animal world. Perhaps the reward is not food; perhaps it is intrinsic in the performance. The work of a specialized dog must be like play or courtship is for humans—it just feels good to do it. Maybe the evolution of dog behavior is driven by the developing form that is rewarded simply by the performance. The pat on the head is extra.

In Part II, I will explore the various mechanisms by which breeds of working dogs evolved. As we will see, the processes of creating working dogs are complex. In some instances we see that the working behavior is simply a behavioral adaptation to the changing economies of humans. In other cases the form of the dog is a product of natural selection operating to adapt the scavenger village dog of Mesolithic peoples to the Neolithic agricultural way of life. Scavengers living within sheep cultures, for example, are dogs that are exposed to different selective forces than are the basic village dogs.

Each of the working breeds is an adaptation to human endeavors by the basic village dog. In many cases humans do influence the direction of the evolving form, but not as much as we may have thought. Breed development was not always artificial selection in the way Darwin envisioned. Breed evolution might actually surprise Darwin, in being more like natural selection than he imagined.

Each chapter in Part II illustrates a different mechanism that changes dogs from commensal to mutual symbionts with people. In Chapter 4, village dogs mature into working and sporting breeds through a series of developmental environments and events that shape their adult *behavioral conformation*. No human-purpose breeding is necessary to achieve animals capable of performing complex tasks. In Chapter 5, sled dogs are examples of a perfect *physical conformation* for accomplishing their task. This physical conformation is achieved by culling nonperformers from the gene pool. Each generation gets better and better. But, what is happening is that the less efficient animals are being culled, rather than any recognition by humans of what characters are being selected for or

what constitutes good conformation. Chapter 6 describes a parallel formation of specialized herding and gun dogs. In this case, artificial selection is for a perfect *behavioral conformation.* A behavioral conformation for
pointers, retrievers, and herders is just as necessary as the physical conformation is for sled dogs. Once the breed has the exact physical or
behavioral conformation, it can perform its task better than any other
species. For example, sled dogs can run marathon distances faster than
any other species. Without the exact behavioral conformation, it is
doubtful that one could train a dog to perform its task satisfactorily. Just
as it would be silly to try to train a dachshund to pull a sled fast, it is similarly problematical to win a herding-dog trial with a golden retriever.

Part II also provides insights into the nature (genetic) versus nurture
(environment) issue. Only dogs with the correct natures can be trained
to work. But even among dogs of the correct nature, most cannot perform unless they are nurtured properly from birth. I will sort through
behavioral differences in dog breeds to show how behaviors are genetically biased and how the environment influences their development.
Why is it that even Erik Zimen, an expert wolf behaviorist, couldn't
train a few wolves to be real sled dogs? Why can't I train wolves to be
herding or guarding dogs? Why can't I train border collies to be guarding dogs? Why are there even some guarding dogs I can't train to be
guarding dogs? Note the plurals. I'm talking about the improbability of
training many specifically natured dogs to be other than what they
were bred for. I've heard from people who claim to have wolves on their
racing team, or whose guardian dog also herds the sheep. I know there
are isolated examples of dogs doing remarkable and uncharacteristic
tasks. But I also understand that many who love their dogs impart
supercanine abilities to them, letting their imaginations overrule their
observations.

CHAPTER 4

Developmental Environments

LIVESTOCK-GUARDING DOGS

An ocean of grass extends westwards from Manchuria to the Hungarian Plain. Over its undulating horizons, mounted nomads moved their flocks on a restless search for food. In winter they sheltered under the lee of mountains from the *buran* or white wind of winter; in the spring they relaxed when the flowers lacquered the ground . . . Their migration was their seasonal ritual, their music the howling of mastiffs, clanging of bells, and pattering of feet. [Bruce Chatwin, *What Am I Doing Here.* New York: Penguin Books, 1990, p. 197.]

Dogs . . . are of the greatest importance to us who feed the woolly flock, for the dog is the guardian of such cattle as lack the means to defend themselves, chiefly sheep and goats. For the wolf is wont to lie in wait for them and we oppose our dogs to him as defenders. [from *Roman Farm Management: The Treatises of Cato and Varro* (circa 150 B.C.) *Done into English, with Notes of Modern Instances, by a Virginia Farmer* (Harrison Fairfax) New York: Macmillan, 1913, p. 247.]

IN THE NEST: SHAPING THE BEHAVIOR

There are several types of sheepdogs. There are the herding dogs, which conduct livestock from one place to another. And there are the livestock-guarding dogs, which cannot herd sheep, are not expected to herd sheep, but are expected to protect sheep from predators such as wolves, coyotes, bears, jackals, baboons, leopards, or any other depredator that pastoralists might encounter. (Note that I use "sheep"—as Cato and Varro used "cattle"—as a synonym for all livestock.)

The herding dogs and the livestock-guarding dogs are an animal behaviorist's dream team of study animals. Two types (breeds) of dogs are raised in the same environment—pastures; they are both selected to respond to the same environmental stimulus—sheep; but they respond in two very different ways. One herds the sheep, the other guards them. Since the two breeds behave differently in the same environment, then we can assume the differences between them are genetic.

Livestock-guarding dogs are probably among the oldest of the working dogs. Obviously, they cannot be older than domestic sheep and goats, which are the first livestock, dating from about eight thousand years ago. At Cato's time (2,150 years ago), livestock-guarding dogs were common and economically important. They show up frequently in ancient writings and pictorial art.

They are probably also the most numerous of all the working dogs. There are millions of them throughout the world. We in the West think of them as rare breeds, such as komondors, kuvasz, or Great Pyrenees. When I started studying livestock-guarding dogs in the 1970s, I found practically nothing written about them in the United States. What there was often turned out to be wrong. Breed books claimed that these big sheepdogs were both guardians and herders. English-speaking peoples tend to think of sheepdogs as collies or herding dogs. Perhaps this confusion was because the English-speaking world had either rid itself of significant predators many centuries ago and had no need for livestock-guarding dogs, or, more likely, the breed-book authors had not really observed closely the behavior of the dogs with the flocks.

In fact, when twenty-four-year-old English biologist Charles Darwin encountered livestock-guarding dogs on his visit to Uruguay in 1833, he was "amused" with what he saw. It was as if he had made an original discovery unique to South America. He didn't seem to realize that the raising and training procedures he reported on were standard the world over (except in the British Isles). He didn't seem to realize that there are, and were in 1833, millions of these dogs, with four or five of them each working for just about every shepherd from Portugal to China, from Russia to South Africa.

But Darwin is such a keen observer and wonderful writer, I will present his discovery in his own words. He manages to capture all the elements of the livestock guardians in just a few sentences.

> While staying at this estancia, I was amused with what I saw and heard of the shepherd-dogs of the country. When riding, it is a common thing to meet a large flock of sheep guarded by one or two dogs, at the distance of some miles from any house or man. I often wondered how so firm a friendship had been established. The method of education consists in separating the puppy, while very young, from the bitch, and in accustoming it to its future companions. An ewe is held three or four times a day for the little thing to suck, and a nest of wool is made for it in the sheep-pen; at no time is it allowed to associate with the other dogs, or with the children of the family. The puppy is, moreover, generally castrated; so that, when grown up, it can scarcely have any feelings in common with the rest of its kind. From this education it has no wish to leave the flock, and just as another dog will defend its master, man, so will these the sheep. [Charles Darwin, *The Voyage of the Beagle,* New York: P. F. Collier and Son, 1909, p. 163.]

Note that he says nothing about their breed, special breeding, selection, or anything to do with genetics. Breed is not an issue. In fact, from Darwin's description one has no idea what the dogs look like or how big they are. My guess is they are not much more than useful village dogs, adapted to the flock. Darwin was such a good reporter, surely he would have said something if these dogs were really big or beautiful or strikingly special in some way.

The message inherent in Darwin's description is, Take any local puppy and raise it properly and you have a decent livestock-guarding dog. Uruguayan shepherds, or any shepherds, for that matter, knew how to achieve good guardians. One hundred and fifty years after Darwin's observations, when biologists Hal Black and Jeffrey Green were trying to teach modern-day ranchers in the western United States how to raise and train livestock-guarding dogs, they reproduced the recipe used by Navajos to develop their flock guardians. They acquired this recipe by observing Navajo sheepdogs. The Navajos were originally taught by Spanish missionaries that the proper way to care for sheep is to raise sheep-guarding dogs with them. The system sounds very much like Darwin's formula.

Raise or place mixed-breed pups in corrals with sheep, lambs, goats, and kids at 4-5 weeks of age. Feed the pups dog food and table scraps. Provide no particular shelters such as dugouts or doghouses (the pups will sleep among the sheep and will dig their own dirt beds). Minimize handling and petting. Show no overt affection. Return pups that stray to the corral (chase them, scold them, toss objects at them). Allow pups to accompany the herds onto the rangeland as age permits. Punish bad behavior such as biting or chasing the sheep or goats, and pulling wool by scolding and spanking. Dispose of dogs that persist in chasing, biting, or killing sheep. [Black and Green, 1985]

In both descriptions the critical factor for achieving the appropriate adult behavior is to start with very young pups and raise them with the target species, without other dogs around. Darwin says, "I often wondered how so firm a friendship had been established." How can a carnivore become a protector of a prey species? Then Darwin answers his own question by describing the essential developmental environment. Instead of the dog being raised in the house, where it becomes trustworthy, attentive to, and protective of its master, it is raised in the barn, where it grows up trustworthy, attentive to, and protective of sheep.

If Darwin had had a modern vocabulary, he might have written: the interspecific social bonding between sheep and dogs depends upon *imprinting* puppies during the *critical period* of socialization, which for dogs is roughly between four and sixteen weeks of age. By paying strict attention to the puppy's developmental environment, one shapes and conditions the dog's adult behavior in such a way that it displays normally intraspecific social behaviors (innate dog-dog behaviors) interspecifically (nurtured dog-sheep interactions). As a result of this rearing environment and imprinting, the dogs cannot display predatory routines toward sheep. We have seen this phenomenon already with village dogs in Pemba, which, being raised with chickens, don't kill them.

Most shepherds don't even realize they are manipulating the dog's behavior, just as most of us have little knowledge of how we change a dog's behaviors by raising it from puppyhood in our home. We buy an eight-week-old puppy, take it home, where it is isolated from other pups, and it grows up with us. People, then, become the dog's social attachment. Livestock-guarding dog pups are born in sheep barns, form their attachments during their first few months of age, and grow up socialized to sheep.

The only difference between the commensal village dogs, which were difficult to lay a hand on, and our pet dogs, or livestock-guarding dogs, is the social environment they were raised in.

Will any breed of dog do? Can you take any dog and start at four weeks and make it into a livestock-guarding dog? No, not really. We raised a retriever according to the recipe, but she never developed the protective attitude that the guardians do, and she never forgot how to retrieve. None of the specialized breeds I discuss in Chapter 6 will make good livestock-guarding dogs. And I will explain why.

Cato the Elder might have been among the first to recognize this distinction, and his advice to farmers over 2,000 years ago is still irrefutably valid today.

> Be careful not to buy a sheep dog from a professional hunter or a butcher, because the one is apt to be lazy about following the flock, while the other is more likely to make after a hare or a deer which it might see, than to tend the sheep. It is better either to buy, from a shepherd, dogs which are accustomed to follow the sheep, or dogs which are without any training at all. While a dog does readily whatever he had been trained to do, his affection is apt to be stronger for the shepherds than for the flock. [p. 249]

Livestock producers who rely on their guarding dogs come to the same conclusion. Breed, shape, and genetics are not as important as the developmental environment. I will modify that statement ever so slightly as I show how the breeds of livestock-guarding dogs evolved. But for the most part, while the dog is in its first few weeks of life, and growing its brain, it is making the cell connections and rearranging them in a specific way, according to the signals that are coming from outside. This development predetermines its adult behavior. In other words, imprinting changes the dog forever.

Austrian ethologist Konrad Lorenz was maybe the first to recognize the importance of this period of primary socialization, which he was able to articulate suitably and which he tested in his now-famous experiments with birds—observations that won him a Nobel Prize. Working with graylag geese, he demonstrated that socialization with another species (namely, with Lorenz himself) during a time-sensitive period of development resulted in the birds' being "imprinted" on him. Geese are a precocious species, walking and swimming within an hour of hatching. In some cases social bonding takes place within *minutes* of birth. In contrast, with Lorenz's jackdaws, an altricial (hatched in a helpless and naked state) species, the bonding was more gradual and occurred later in development. But if the bonding was done correctly, both species would prefer the company of their "foster parent."

Dogs have altricial young more like the jackdaws than the geese. Dogs are born without eye and ear function. Their survival, like that of the jackdaws, is dependent on parental care. They can't make it on their own. The concept of critical period resulted from the differences Lorenz recorded between the goslings, which can make it on their own and which form their social attachments immediately upon being born, and the jackdaws (and dogs), which need several weeks to form theirs.

Wolves, coyotes, jackals, and dogs all begin to form their social allegiances after their eyes open at thirteen days. With the onset of sensory functions, they have the capacity to form social relationships. By the time dogs are sixteen weeks old the window of social opportunity is greatly diminished or even closed. If they haven't seen sheep or people during that period, they will be forever shy of them. Wolves are different from dogs in that although they begin their social development at thirteen days, it is greatly accelerated and is closing rapidly by nineteen days. One of the reasons dogs can be so much tamer than wolves is a consequence of the much longer period during which dogs can form new attachments.

The period roughly between two and sixteen weeks, called the "critical period for social development," was originally described for dogs in a 1950 paper by John Paul Scott and Mary Vesta Marston, resulting from the notable studies at Bar Harbor, Maine. Critical period simply means that during this time, the pup is predisposed to and has the greatest capacity to learn particular social skills. It is in this period when dominance hierarchies are formed and dogs learn and practice their submissive behaviors. They learn to beg for food, whom to beg from, and how to turn begging into social greetings. They learn what species they belong to.

At sixteen weeks the social learning window closes. After that the dog has very poor abilities to develop or change its social skills. Essen-

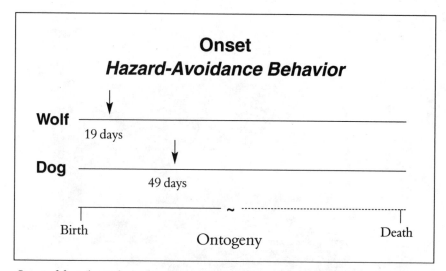

Onset of fear (hazard-avoidance) motor patterns in wolf and dog. The onset of fear motor patterns occurs at about nineteen days in wolves. If wolves haven't been introduced to people by then, it is doubtful they can ever be tamed. Dogs, on the other hand, don't have the onset of fear until the sixth to eighth week. It is much easier to make a pet out of a dog.

tially, at sixteen weeks, the dog's social personality is set for life. If a dog is shy of people at sixteen weeks then it will be shy the rest of its life. Can it learn not to be shy with intense training? It certainly could make some progress but it will always have a social "accent."

Is there any variation in the closing of the social window? Of course. Individuals will differ within a breed, and the average for each breed will differ. In fact, at Wolf Park the puppy socializers remove the puppies from the den before their eyes open. Because wolves' social abilities mature so rapidly, the human puppy parents have to spend twenty-four-hour days with them from their thirteenth to nineteenth days. And that includes feeding, cleaning, and playing with them. You can see why Mesolithic puppy snatchers would have had such a hard time taming and training wolf pups. Removing pups from mom and taking care of them during that neonatal period are extremely difficult and time-consuming. Just bottle-feeding them is a chore.

When a future livestock-guarding dog is raised among sheep for its first sixteen weeks, then for the rest of its life it treats sheep as its primary social companions—it is imprinted on them in a way similar to

Dominant and submissive motor patterns emerge during the critical period of social development. (K. Doktor-Sargent)

Dog submissive behavior displayed to sheep. When a pup begins to show dog social motor patterns toward sheep, one can conclude it has successfully bonded with them. This Anatolian shepherd dog, displaying a classic dog submissive posture to these sheep, is developing into a fine guarding dog. (Photo by Jay Lorenz)

the imprinting of Lorenz's geese. The adult guardian dog follows sheep, greets them, and responds to their signals by showing dominance and submissive behaviors interspecifically. Sometimes these dogs even get sexually involved with sheep.

Some people say that the dog thinks it is a sheep, but that is probably wrong. It "knows" it is a dog. The behaviors it directs toward the sheep are dog social behaviors, not sheep social behaviors. If it thought it was a sheep it would display sheep social behaviors, which are very different. When dogs threaten they show their teeth and growl, while a sheep threatening a dog stamps its front foot. Growling as a social response is genetic, a dog characteristic. Dogs do not normally growl at their prey. Growling is directed at animals they need to communicate with. Thus, growling at sheep is a good sign that the dog has developed a social relationship with sheep during the critical period. Who they socialize with is learned, but dogs can learn this lesson only during the critical period, which is genetically timed.

When I talk to my dog I'm communicating by means of human social behaviors. I don't think I'm a dog. But there is a part of me that thinks my dog is human. I even think my dog knows what I'm talking about. I watched a Portuguese shepherd yelling in perfect Portuguese at his goat, which was browsing his neighbor's garden. The goat must have understood Portuguese because it hastily got out. It seems that livestock-guarding dogs, in the same way, think sheep are doglike and understand dog language. When I feed my dog, the sheep try to steal as much dog food as they can, and the dog growls at them, as if they could understand what it meant.

Typically, Old World shepherd dogs spend their first sixteen weeks with one or two littermates, a few adult dogs, including their mother, three hundred or so sheep, and a shepherd. The Italians have a word for this social triangle. It is called a *morra*. After the sixteen weeks are over, the dog has been physically shaped and behaviorally molded in such a way that it "needs" to spend the rest of its life with the morra. One would never think to purchase such a dog and take it back to a city apartment as a pet. It probably would be very uncomfortable and could never make a total adjustment to the nonsheep environment. Taking a wolf pup from the den at four to six weeks would produce a similar unsatisfactory result.

The practical aspects of critical period contribute much to our relationships with dogs. In fact, the first and most important aspect of creating a mutual relationship with dogs is not genetic at all, but rather the development of puppies in the environment they are expected to perform in as adults. Unfortunately, the critical period is often poorly understood, even by trainers whose job it is to shape a dog's behavior for a specific use as an adult. For example, a pervasive view describes the social behavior within a pack of wolves as genetic. Because of this, the reasoning of dog trainers goes: dogs are descended from wolves and wolves form packs, and therefore dogs understand wolf-pack behavior and should respond to the trainer as "alpha," or dominant, in its life.

But *is* wolf-pack behavior genetic? Not really. Pack behaviors, like all behavior, are epigenetic—above the genes. They are a result of behaviors learned during the critical period. Pack behavior is just one of many social options available to wolves. If dogs don't develop pack social behavior during their critical period, there is no sense in trying to simulate pack leadership after that social window closes. Pack behaviors are much more complicated than just hierarchies of social status. They are learned through social play and care-soliciting behaviors during the juvenile period. A trainer who pretends to be the alpha leader of a wolf pack—say, by turning a dog over onto its back and getting down and growling at its throat—is intimidating the dog, no doubt. But to a dog, the message is not what the trainer thinks it is. Teaching and learning are seldom facilitated by intimidation. A dog doesn't learn how to sit from a trainer who intimidates it, simply because the coercion diverts the dog's attention away from the task and toward its social status. An alpha wolf is not trying to teach a pack member anything, especially to sit. The fact that so many believe the wolf-pack homology, and use it in training a dog, is really a testament to how little is understood about canine behavioral development.

Critical period for social behavior sounds like magic. Something permanent is actually happening in the dog's brain that causes it to become essentially unalterable after the period is over. For some reason, what is learned, and when it can be learned, is limited to that time period. Once "learned," the behavior cannot, easily or completely, be unlearned. Given how much we do know about teaching and learning, it would seem that we could teach the dog to behave differently. But the dog doesn't appear able to learn it. Proverbially, people do know this: you can't teach an old dog new tricks. But do they know why?

Some permanent change must be taking place during the critical period. It really does look like a dog that is socialized with sheep is wired differently than that same set of genes growing up in a village without sheep. Could it be that the ability to learn is a genetic response to the environment? Could learning be genetic?

There is an important essential here. Early experience is vital not because it is the first learning, but rather because it affects the brain's development. A while back, I was training dogs to be livestock-guarding dogs—I *thought*. I thought the dogs were learning to be flock guardians. The critical period closed on the sixteenth week or thereabouts, so, I reasoned, if we got our dogs out with sheep any time before sixteen weeks we would be fine. But by keeping our pups with sheep, we were not teaching them to be flock guardians. Instead, the young dogs were "growing" their flock-guardian-behavior brains.

Brains grow, just like legs or any other body part. Legs not only *can* walk but they *must* walk in order to grow properly. Legs that do not walk while they are growing (critical period) wither and become useless. The same is true of brains. Brains grow in two ways: they get bigger, and they change shape. How much they grow and which way they change shape depends on the kinds of environmental stimulation they receive during their first sixteen weeks.

The period of the most rapid brain growth is also coincidental with the critical period for social development. A day-old livestock-guarding puppy has a brain volume of about ten cubic centimeters. That is about the size of the very end of my little finger. By the time the pup is weaned at eight weeks, it is sixty cubic centimeters. By the time it is sixteen weeks, its brain is eighty cubic centimeters and rapidly approaching full size at a little over one hundred centimeters. Tom, one of our original and favorite Italian sheepdogs, grew to a hundred pounds with a brain volume of a hundred and six cubic centimeters.

At birth a puppy has essentially all the brain cells it is ever going to have during its whole life.

If the puppy brain has essentially the same number of cells as the adult brain, how can it grow ten times bigger? The answer is that brain growth is almost entirely in the connections between the cells. Of all the brain cells present at birth, a huge number are not connected or wired together. What takes place during puppy development is the wiring pattern of the nerve cells. Some nerves make their connections spontaneously, driven by internal signals. Some nerves actually "look" for a

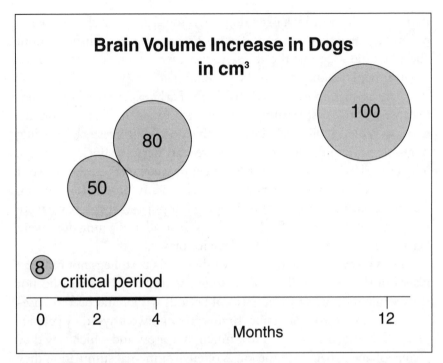

Brain growth chart. Most of the growth in a dog's brain is during the critical period of social development. After growth stops, it is difficult to change the wiring.

muscle to attach to. Other connections are motivated by external signals. External to the brain, that is. For example, the eye tells the brain how many cells it needs to have in order to run the eyeball. Big eyes need more cells than small eyes, and thus animals with big eyes tell their brain to connect up a greater number of cells for eye function.

It is not only the size of the eye to which the brain must accommodate, but also the activity of the eye. The brain accommodates to the eye by growing the appropriate connections for both its size and its activity. The brain of a puppy raised in the dark doesn't make as many connections. A puppy raised in the dark has a smaller brain than one raised normally. A puppy that is raised in an impoverished environment has a smaller brain. It has the same number of cells, but not as many get wired together. Experiments with kittens showed that animals raised with horizontally striped glasses during development of the eyes cannot, as adults, see in the vertical plane. They walk into table legs as if they can't see them.

The onset of other sensory functions, too, happens in the critical period. Animal behaviorist Ed Bailey notes the importance of olfactory imprinting for establishing a bond with a young pup. He advises prospective owners to visit the litter from five weeks of age on, handling the chosen pup and allowing it to imprint on specific smells. Darwin in Uruguay noted that the guarding-dog pups were given "a nest of wool . . . in the sheep pen." From birth, they lived amid the feel, sight, sound, smell, and taste of sheep. Among gun-dog breeders, it is known that young pups can easily be accustomed to the sound of a gun firing nearby, if they are exposed to this and other loud noises before the onset of fear.

At this point, it should be clear that the sight, smells, and sounds of sheep influence the growth of a brain in a manner distinct from the influences on the brain of a puppy that grows up in a city apartment. If the puppy doesn't see sheep until it is eight weeks old, it has a differently shaped brain than one that sees sheep from four weeks old. At sixteen weeks old, almost all the connections have been made, and the brain is about to stop its major growing. A puppy that sees sheep for the first time at sixteen weeks can make some tiny growth adjustment to those environmental signals, but not as many.

It is complicated and would take massive computers to figure out all the possible variations. But at the same time it is conceptually relatively easy to visualize what is happening. Why is it that no two animals truly look alike or think alike? Simply because no two animals can grow up in identical environments. No two animals can occupy the same space at the same time, which means they cannot receive identical environmental signals, and therefore they do not get wired up identically. As a result, they do not have the same capacity to think and behave. Even if the impoverished pup gets taken to an enriched environment as an adult, it cannot learn to cope with that environment because it does not have the necessary cell connections. Once the dog gets to sixteen weeks it has made (or has not made) just about all the social connections it is ever going to make.

Understanding brain growth should dispel the nature/nurture controversy once and for all. It is never, ever either nature *or* nurture, but always both at the same time. But liver cells make more liver cells because that is the environment they respond to. Behavior is always epigenetic—above the genes—an interaction between the genes and the developmental environment. It is a synergism of gene products and the climate, producing a unique organism. No two look or behave alike.

Development of the growing brain is a cascade of billions of events that are internally (nature) and externally (nurture) motivated. To say, however, that the cascade of growth events is genetic is to miss the point.

In 1976, when I started studying livestock-guarding dogs, I assumed that since the behavior of guardian dogs was so dissimilar to that of border collies, then guardian dogs must be preprogrammed to be livestock guardians. I thought livestock-guarding-dog behavior was genetic. That was what everybody was telling me. These breeds, everyone said, had been selected to guard sheep. We were also constantly asked, "Which breed is best?" The implication was, which breed has the best genes?

At the same time, during our research program with the guardians, we received many telephone calls from producers with the following complaint. The caller had purchased an older pup—say, a four-month-old Pyrenean from a breeder who told him it was a traditional livestock guardian—and he couldn't get the dog to stay with the sheep even though the dog was from "good" breeding stock. Our first question to him: Was the dog in with sheep for its first four months? No? Then it had the wrong brain shape. You can't satisfactorily teach a dog a new social trick.

Raising puppies, and especially raising them for special jobs as adults, requires attention to detail. When people raise pups as pets, they often get them at about eight weeks old, take them home, feed and cuddle them, housebreak them, take them for walks, and play with them. What they are doing (and they're usually not aware of it) is providing specialized brain-growing conditions that shape the dog's future behavior. If I were buying a puppy for a pet, I would check its early environment and make sure it wasn't raised in a kennel or in the laundry with only its mother and littermates for immediate company during that first eight weeks. I'd be very suspicious of a department-store dog that was twelve weeks old, wondering if the dog had time left to grow the brain I was looking for. I'd also suspect that if I locked the pup up in the house alone each day while I went to work, I'd get a small-brained dog without enough connections to be a good social companion.

Critical period needs to be explored even further. It implies so much more than simply animal-to-animal socialization. For example, one reason Konrad Lorenz's geese stopped imprinting shortly after birth was because they experienced the onset of a fear response. Fear is a threshold response, meaning that a stimulus has to exceed a certain level before it provokes a response. Take any signal. The animal could respond to the signal by approaching it curiously, or the animal could try to escape or avoid whatever produced the signal. Take a sound. The same sound could be loud or it could be barely perceptible, depending on the ear's development. As the sense organ begins to function, what might be a loud sound to an adult is a soft sound to an immature ear. Animals avoid loud sounds and could ignore or investigate soft sounds. They could habituate to sounds that were continuous. Fear is in great part an avoidance of novelty.

Before the onset of fear responses, animals do not show fear to novel shapes and sounds. For a newborn puppy, everything is novel. But after the onset of fear, new novel shapes and sounds cause avoidance behaviors—call them hazard-avoidance behaviors. Gun-dog trainers expose their pups to gunshots before the puppies grow the "fear" portions of their brains. Shooting guns around puppies for their first six weeks grows a brain that expects those sounds from the environment. Gunshots become normal. But if the gunshots are not introduced until after the onset of fear (which in this case might simply be the threshhold where the sound is perceived as loud), the dog will perceive them as hazards to be avoided. A gun-shy dog is not much use on the hunt. And as always, once you have a gun-shy dog, there is not much that can be done about it. (That is not to deny that some dogs are more sensitive than others, regardless of the environment.)

Fear turns on at different times in different breeds, and even among individuals within a breed, due to individual development rates. Six to eight weeks is an average age for one breed to display fear responses, while the next might not display them until eight to ten weeks.

Each behavioral system—fear, submission, investigation, play—has its own rate of development, and varies among breeds. Each is dependent on glandular development and hormone secretions, as well as motor coordination and sensory perception. And each feeds back on the puppy to change not only the shape of the brain but the shape of all the other developing organs. And after that each new signal works on that new shape, changing it in a novel way. And so on. The bones of active puppies are a different shape than those of inactive pups. One can change the growth rate of glands by exercising them, and thus change the threshold timing of specific behaviors.

During the critical period of socialization, border collies and some bird dogs show the onset of predatory behaviors. Ten-week-old border collies begin to eye and stalk objects in their environment. They then incorporate eye-stalk-chase games into their play routines with other border collies. Thus, part of their social play has predator-versus-prey components. Sheep are very sensitive to predator-versus-prey behaviors and avoid animals that display eye-stalk games.

If livestock-guarding dogs ever display predatory behaviors, they generally do not appear until the dog is five or six months old. By then, their window of social development is closed. Therefore they cannot integrate those predatory behaviors into their social play. That in itself makes their breed personality very different from collies.

The basic predatory behaviors (called predatory motor patterns) of a dog are:

orient/eye/stalk/chase/grab-bite/kill-bite/dissect.

Not all breeds of dogs have a complete set of predatory motor patterns. In fact, one of the clues about the status of the village scavenger on Pemba is that this dog didn't display any, at least not to village livestock or chickens.

Dogs and carnivores in general don't show kill behaviors to animals they grow up with, or individual animals they know. Ethologist Paul Leyhausen, in his book *Motivations of Human and Animal Behavior,* described an example of this phenomenon. He had a wild golden cat (*Profelis temmincki*), to which he fed rats. One rat avoided being caught and hid under the cat's bed. That rat became "friends" with the cat. The cat ate each new rat but not its friend—until at four months the rat was removed, to be put back, much larger, three months later. The friendship was over.

We had five chickens that our Alaskan husky Sitka "knew." She never touched them. They got old and I added two more. Sitka killed and ate both new ones and still didn't touch the old ones. Breeders of working foxhounds know how this works, and place their puppies out to be raised by farmers known as puppy walkers. Since these hounds don't develop their hunting instincts until after the time during which they have been socialized with farm animals, the adult foxhounds, thus raised, can chase a fox through the farmyard and never look at a chicken, even one they don't know. The knowledge and use of critical period by foxhound breeders greatly improves hunter-farmer relationships.

The very best livestock-guarding dogs never develop any predatory motor patterns. The less than the best (which is practically all of them) display one or two predatory motor patterns (most likely chase and grab-bite), but very weakly.

What that means is that many an imperfect dog makes a very good sheep guardian if the owner has paid strict attention to the development of the animal during the critical period. But with other animals the good guardian may not be so circumspect. Some will chase and sometimes kill wildlife. In Texas, one of our Maremmano-Abruzzeses, named Dolly, was a wonderful guarding dog—trustworthy, attentive, and pro-

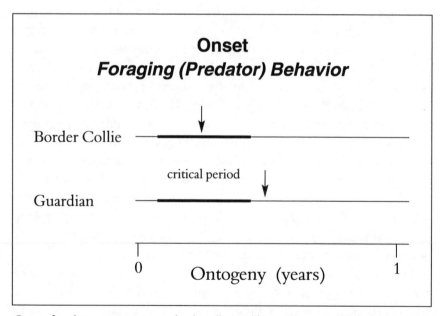

Onset of predatory motor patterns: border collies and livestock-guarding dogs. Border collies have an early onset of predatory motor patterns that they incorporate into their social play. Predatory behaviors of livestock guardians onset after the social window has closed. Their play behavior is not as rich in behavioral elements as that of border collies.

tective with mohair goats. But she would catch and kill a rabbit, and then carry it around for days until it wore out. Then she would get another. She had the chase and kill motor patterns, but not dissect. Therefore she could kill animals she hadn't been socialized with, but she couldn't eat them because she was missing one of the essential elements of predatory behavior. The rancher's interpretation was that Dolly just needed a soft pillow, for he often observed her resting her head on the rabbit carcass.

Just as the social motor patterns have a critical period, so do the predatory ones. Each of the patterns of eye, stalk, chase, grab-bite, kill-bite, and dissect emerge independently from one another during their own critical subperiod. One day you observe the kitten eyeing the ball of yarn. Days later it does the chase and the pounce. Eventually it ceases displaying the behavior to yarn and turns its attention to mice. It appears that the animal is learning to kill, or practicing to be an adult, by playing the eye-stalk games with the yarn. Actually, something much more elaborate and serious is going on.

Motor patterns need to be reinforced the first time they appear. Not all of them, but many, if they are not reinforced during the onset period, drop out of the repertoire, never to appear again. If the animal performs the new pattern then it will persist. Presumably the behavior is "growing," and if it is not stimulated it does not grow properly. Suckling in newborn puppies is a good example. The suckling motor patterns, sometimes called the suckling reflex, turn on in most mammals slightly before birth. If the puppy doesn't suckle within a few minutes after birth, the behavior extinguishes, presumably because it does not grow the proper nerve connections. The puppy that does not suckle within a few minutes after being born loses the ability to perform the sequence. It cannot be taught to do it later. Not only can you not teach an animal to suckle, you cannot teach one to chew and swallow. These are innate motor patterns that develop or onset later.

I've seen cases of livestock-guarding dogs that showed the onset of the chase motor pattern, but if they were removed immediately from the sheep pasture, the behavior dropped out of the repertoire, never to appear again. I've also seen cases of livestock-guarding dogs, with the onset of chase, running at the sheep. My sheep, which had a long history of being socialized with dogs, didn't run in response to the chase. If the sheep don't run, the dog can't chase them. Even though the dog has the potential to chase, it needs the environmental signal in order to display the behavior.

The layering and interacting of developmental events that produce an adult working dog are precisely unfathomable. The complexity of the developing dog's behavior should remind the reader how passé is the nature-versus-nurture controversy. Although it was once a compelling question for behaviorists, scientists now understand why nature cannot ever be separated from nurture. When we look at the critical period for social development, we realize that the genetic nature of the dog is being shaped by the environment in which it is growing up. If there is no environmental stimulation, there is no epigenetic response.

Most good working-dog people make their own dogs. I might buy myself a border collie pup, but that is when my work begins. I have to shape the behavior of that pup from an early age in order for it to be the herding dog I want.

With livestock-guarding dogs, the shaping often comes naturally. The dog is born in a sheep culture. The rest is natural. To the people who call me and ask,"What breed of dog is best?" I reply, "None of them will work at all if they are not raised with sheep during the critical period. The only thing you are looking for is a breed of dog that has a weak tendency to show the predatory motor patterns, a breed where the predatory motor patterns drop out if not reinforced, and dogs four or five weeks old that can be socialized properly. A village scavenger can have all those qualities. Go capture a village pup and raise it with sheep."

THE TRANSHUMANCE: DISTRIBUTING AND MIXING GENES

The behavior of a good working sheepdog is the interaction of a number of environmental, genetic, and epigenetic events, occurring during development. The behavior of the original livestock guardians is not predicated on any genetic evolution of these sheepdogs. And yet we now think of them as breeds—breeds selected to be large, white, and protective. Often, when livestock-guarding dogs are discussed, the topic hinges on which breed is best. Our research at Hampshire College focused on breeds, and we compared the success rates among several hundred each of Italian Maremmano-Abruzzeses, Anatolian shepherds, and Yugoslavian Šarplaninacs. We also observed a smaller number of Tibetan mastiffs, Portuguese Castro Laboreiros, Pyrenean mountain dogs, komondors, kuvasz, pulis, and two Transcaucasian ovčarkas, to assess how they worked on American farms and ranches.

Why, if their adult behavior is shaped by their developmental environment, do these livestock dogs have distinctive shapes also supposedly adapted to guarding livestock? We must explore how the ancient dogs metamorphosed from village dogs growing up among shepherds and their flocks to the special shapes of today's great breeds.

The Roman statesman Cato the Elder describes in discerning detail the appearance of flock guardians in 2150 B.P. as follows.

> . . . handsome, of good size, with black or tawny eyes: a symmetrical nose: lips blackish or ruddy, neither drawn back above nor hanging underneath: a short muzzle, showing two teeth on either side, those of the lower jaw projecting a little, those above rather straight and not so apparent, and the other teeth, which are covered by the lips, very sharp: a large head, ears large and turned over: a thick crest and neck: long joints: straight legs, rather bowed than knock-kneed: feet large and well developed, so that in walking they may spread out: toes slightly splayed: claws hard and curved: the pad of the foot neither horny nor hard, but as it were puffed and soft: short-coupled: a back bone neither projecting nor roached: a heavy tail: a deep bark, and wide, gaping chops.

I suppose this might have been the first breed standard ever written. Its overall message is that the dog should be well built. It doesn't say big or fierce but rather that the dog should be nicely shaped with good pigment. This description implies that shepherds, by ancient Roman times, were paying attention to the conformation of their flock dogs— the conformation present in the best working dogs. Cato added one more trait: "The colour to be preferred is white because it gives the dog a lion like aspect in the dark."

To state that the preferred color is white implies that there are good working dogs that are not white.

I suppose it could be interpreted from Cato's description that the ancient Romans were breeding livestock-guarding dogs. Unless one reads carefully, one might think the Roman ranchers were practicing artificial selection in the Darwinian fashion of sorting among their best dogs and selecting traits such as the white coat color and breeding together those dogs with the desired characteristics. We might assume that since the Italian Maremmano-Abruzzese is white, people have been breeding these dogs since ancient times, preserving ancient traits generation after generation. We might be tempted to think the Maremmano-Abruzzese is an ancient breed.

I don't think that is so. I don't think that is the way breeds of livestock-guarding dogs evolved. And I don't think any breed is ancient in the sense that it has been sexually isolated since Cato's time and is directly descended from ancient ancestors. I have no doubt that some Romans had dogs that were uniform in traits and looked like breeds, and maybe they even looked like modern breeds. I'm sure there were distinctive livestock-guarding dogs and distinctive hunting dogs. And I don't doubt that there were a few Roman aristocrats who kept kennels, and kennel personnel who arranged matings between individual dogs. Just as now there are Walker coonhounds, there must have been Caesarian spaniels.

But it is a long way from some rich Roman kennel to the millions of shepherds and their multimillions of working sheepdogs in remote mountain pastures. Survival as shepherd dogs depends on the phenomenon of the "transhumance." Every spring and every fall, shepherds move their sheep to where the grass is fresh and plentiful. They wander vast rangeland pastures and along mountain trails. While traveling, they don't have the facilities to isolate their female dogs in estrus. A female in heat is receptive to males for about two weeks. Think of walking around one of these huge grasslands every day, camping each night in a different location, and trying to prevent a female from getting bred by every male within olfactory range. Modern shepherds in these same sheep-producing nations still don't have any system for protecting a female in heat. It is hard to imagine that in Cato's time or in the early Neolithic period, shepherds could have had control over their dogs' reproductive urges. When I think of the problems I have, kennels and all, and the mistakes I've made arranging matings, it is just impossible for me to believe that any of these "breeds" are the result of sexual isolation over centuries.

Dogs in heat inevitably get bred, first by the dogs they socialize with daily, then by the other dogs of the *morra*. Then by dogs from the neighboring morras, and perhaps by all the males from the shepherd's village.

It is a volunteer program. Sometimes a male, or a group of males from the shepherd's flock, will defend the female from other males. Thus the female is most likely to get bred by members of her own morra. But "most likely" is the pivotal term.

All the eligible dogs of the flock and the surrounding flocks and all the dogs of the village can volunteer. The resulting litter could conceivably have as many fathers as it has pups. The dogs closest to the female are most likely sheepdogs, and so even if there are many fathers, they are most likely all neighboring sheepdogs. Thus, all the pups are most likely pure sheepdogs, in that all the volunteers are being tolerated by humans on the sheep range.

By the way, this is the healthiest breeding system possible. Any male carrying deleterious genes does not threaten the whole litter. Random matings among a population of animals that is well adapted to the habitat is a better way to maintain the appropriate gene frequency. And a litter with many fathers should at the same time maximize the variation within the litter upon which natural selection can operate.

Since the livestock-guarding dogs breed themselves, why don't they end up looking like village dogs? In fact, in most of the sheep pastures of the world they are still scavengers and have those village-dog traits of size, shape, color, and behavior. But what has to be realized is that the selective forces that determine the size of the Pemba dogs are not the same selective forces operating on dogs in mountain pastures in Nepal. Even though they are still scavengers, and largely responsible for their own reproduction, livestock-guarding dogs are commonly up to three times bigger than Pemba dogs.

The sheepdogs of the Damara people in Angola and Namibia are not three times bigger, but rather indistinguishable in size from the local village dogs. The sheepdogs of Eurasia tend to be larger than sheepdogs of Africa. Biologists have long known there is a tendency for members of a species to get larger as they move away from the tropics toward the poles. They also get larger as they go higher in altitude. The shepherd dogs of Eurasia live farther north and in higher pastures than the ones in Africa. One might think it was just natural for Tibetan mastiffs to be considerably larger than Pemban dogs—because the latter would freeze to death in the alpine environment.

Thus we must consider that regional variation might be the product of natural selection and not necessarily the result of purposeful selection by people. It is often assumed, when one finds a population of dogs with a high frequency of some attribute such as size or color, that it is because people are selecting for those attributes. But we need to keep in mind that regionally, a population of dogs can be the product of founder effects (Chapter 3), that is, they descended from only a few ancestors. And, further increasing their local uniformity of appearance, they are a product of exposure to local selective forces.

The simplest way to refine a natural breed is by postzygotic selection. This simply means the shepherds cull what they don't like and care for and support what they do like. If a dog chases sheep, they kill it. If a dog doesn't stay with the sheep, it gets lost. If it gets lost, it doesn't have a good chance of breeding with the shepherds' other dogs. If it gets lost, it can't collect the shepherds' and the sheep's wastes.

If a Neolithic shepherd notes that a likable dog is perhaps performing some service such as protecting the sheep from wolves, then the shepherd might share a meal or direct some of the surplus or waste to the preferred dog. All this gives the good dog a better chance of survival and leaving genes to the next generation.

But, take note, it was not bred on purpose to be a good working dog. Nor was it trained to be a good working dog. In fact, the real reason it is a good dog is that it grew up during its critical period with the sheep and shepherd. Its working behavior is partly an accident of the environment into which it was born.

The processes of the natural forces operating to shape the physical attributes of livestock-guarding dogs in the direction of "breeds" are fascinating. Like all natural selection, survival depends on feed, reproduction, and hazard avoidance. Long, long after Cato, I saw my first working livestock guardians in Macedonia, just west of Skopje. Sheep and dogs were on their annual spring migration to the Šarplanina Mountains. I followed a flock of more than half a million sheep walking from lowland winter pastures in Greece up to summer pastures along the Albanian-Macedonian border. I sat at one road crossing for fifteen days with a young Macedonian I'd met through the American consulate there, and watched thousands of sheep, shepherds, and dogs wend their way north. At first the dogs were hard to spot, because they were sheep sized, sheep shaped, sheep colored, and had long, woolly tails very like those of the sheep. The dogs also behaved like the sheep, plodding along with their heads down, mile after mile.

Eventually I followed flocks and their dogs in much of Mediterranean Europe. From west to east, every country has one or more regional variations (or, nowadays, they are called breeds) of guardian dogs. Portugal has at least three—the Estrela mountain dog, the Castro Laboreiro, and the rafeiro do Alentejo. Spain has the mastino español and claims some portion of the Pyrenean mountain dog, which is normally thought of as a French breed.

The Italians have one breed, the Maremmano-Abruzzese, although every so often some group wants to divide it into regional varieties such as the Maremmano and the Abruzzese mastiff. Dog reference books a few years ago designated two Italian sheep-guarding dogs—the Maremmano, with a shorter coat, and the Abruzzese, with a longer body. Then someone noticed that the Maremmano appeared only during the cold months, when the sheep were herded to winter pasture in coastal Tuscany (*mare-mma,* the "pasture by the sea"), and the Abruzzese appeared only during the summer, in the mountain pastures of the Abruzzi Mountains along the central Apennines.

The Yugoslavian Šarplaninac dog is named for the Šarplanina Mountains in what is now southwestern Macedonia and eastern Albania. In the winter, when they are in pasture in Thessalía, they are called Greek shepherd dogs. What starts to be apparent is that each population of dogs has several names, based on where it is located when it is seen.

Most of the sheepdogs in Slavic countries are named ovčarka, or "sheepdog," stemming from the Slavic word for sheep, *ovtsa.* Thus, the mid-Asiatic ovčarka, the (north) Caucasian ovčarka, the south Russian ovčarka, the Transcaucasian ovčarka, the Polish ovčarka, or Tatra mountain dog, for a start. The Hungarians have komondors and kuvasz. Turkish guardians, too, have many varieties, including the Sivas, the Kangal, and the Kars, or, generally, the Anatolian shepherd dog (of the Anatolian Plateau). When political boundaries change, the name of the variety changes, at least in the outside world. The Transcaucasian ovčarkas we bought a few years ago are now called Turkish Kars shepherd dogs. Although I haven't studied them all, I've seen pictures of similar sheepdogs working in Romania, Bulgaria, Georgia, Iran, Iraq, Lebanon, and across into Afghanistan, Tibet, Nepal, and in the Gobi portion of China. Just about every traditional sheep culture employs a "breed" of livestock-guarding dog. A recent compilation for KORA, a Swiss carnivore conservation and management program, named twenty-six countries across Europe and all the way to Tibet with at least forty-eight types of guardians. Their ("breed") names are usually variations on "sheepdog of the (nearby) mountains (or region)."

The modern trend is to ascribe breed status to the dogs of different ethnic groups or national areas. With the current interest in sheepdog breeds and ethnicity, there are more breeds of dogs identified than there are sheep cultures to develop them. It seems that with present-day hobbyists there is virtue in discovering the "names" of ancient "breeds." Within new national boundaries or regions, hobbyists develop rationales (and controversies) about which are the real, the pure, the original breeds. For example, in Turkey recently, some agencies and some non-Turkish hobbyists have initiated efforts to recognize their "breeds" by color variation. The ubiquitous Anatolian shepherd dog, known also as the karabash, which means "black head," has a white variation. Within the past few years this variation of the karabash has gained value abroad, and is becoming one of those newly discovered "breeds"—the akbash, meaning "white head."

Many mountain shepherd dogs are being adopted by breeders, and subsequently registered with national and international kennel clubs. Travelers, diplomats, military people, and expatriates will import a few dogs and start a breed club. Usually they will bring back a pair and find someone else who has also brought back a dog. They form a club, write a newsletter, keep records for several generations, and then apply to a national kennel club, like the American Kennel Club, for the dogs' recognition as a registered pure breed. The membership often disagree on the particulars of what the breed should look like and who has the "real" ones. Then they split into two groups and even change the breed name of the dog.

But how should a breed be identified? Years ago, I asked a shepherd in Portugal, "Is this an Estrela mountain dog?" and he replied, "Are these the Estrela Mountains?" And, voilà, a new breed is discovered! Now when I visit these mountains, even the shepherds are convinced that there is something unique about Estrela mountain dogs.

Once these working dogs acquire identification as a breed, they become the focus of national pride and gain value for their potential in the kennel-club market. Fanciers assert that their breed is best. It is, to me, a kind of master-race mentality of asserting genetic superiority of a regional variety. I have a good friend who claims that the Maremmano-

Abruzzese has to be the best breed of livestock-guarding dog because the Renaissance started in Italy.

It is, or should be, all in fun. But pure breeds are artificial constructs of breed clubs, and the bottom line is money and pride. The dogs living on a mountainside are not a true, pure, kennel-club-defined breed until someone takes a number of them and sexually isolates them from all other dogs. Since shepherds did and do not have the ability to isolate their females in heat, they could not have fashioned a breed, in the AKC sense of the word. However, because of environmental stresses, founder effects, and preferential postzygotic selection, their dogs might have taken on regional characteristics.

A breed standard is a description of a genetic variant, produced artificially by people who have isolated a smaller population from the rest of the species. International club recognition of a breed does not really pivot on biological rules for forming a breed or race of dogs. The terms *breed, race,* and *subspecies* all have similar biological definitions. They all assume that differences in gene (allele) frequency are not randomly distributed over the earth's surface, and that this nonrandomness is dependent on some selection process. *Subspecies* implies that geographic adaptation has taken place which gives a species regional, subspecific characteristics. Finding differences in the coat color of a species on different continents, for example, leads biologists to assume that the color is an adaptation to local conditions. When such variation is found in people, the term *race* is used, rather than *subspecies*. In animals, *race* assumes racial characteristics that grade from one region to the next and have no clear boundaries. *Breed* is a term used mostly for domestic animals. It implies the capture and penning of animals, breeding programs, and arranged matings. It connotes sexual isolation that is achieved not by regional separation or natural selection, but by separation by people and subsequent artificial selection for a preferred trait. To choose white because it is somehow pure, or a corded coat because it is unusual, is capricious. To preserve the regional livestock-dog characters in the captive environment of the breeder is capricious, in that the adaptations to the mountains are not necessarily adaptations to a civilized environment.

As I watched that huge transhumant migration from Greece to Macedonia in the early spring of 1977, I was witnessing a very important mechanism in dog evolution. Transhumance is the seasonal migration of pastoralists or nomads from their winter pastures at low altitudes to summer grazing higher up, or between lower and upper latitudes. I was tracking a form of livestock management that has been going on for many centuries, surely even before the time of Cato the Elder, over two thousand years ago. Stonehenge and other such monuments, built four thousand years ago, may have been calendars designed to tell shepherds when it was time to start for summer grazing. The Bible is in part a history of tribes of shepherds migrating to and from faraway places with their flocks. The Ice Man discovered melting out of his four-thousand-year-old grave in the Italian-Austrian Alps in 1991 might well have been a shepherd caught in a late-spring storm.

The Macedonian migration in 1977 consisted of half a million sheep walking three hundred miles from the lowland Greek wintering grounds to the summer pastures in the Šarplanina Mountains. And then they walked back to Greece in the fall. That is like walking from New York to Washington, D.C., and back every year. For most of us the thought of walking from New York to Washington even once in a lifetime is a little awesome. But thousands of Eurasian shepherds have walked those distances twice a year ever since they were boys.

Understanding the transhumance is critical to understanding "breeds" of livestock-guarding dogs. The transhumance has always had a direct effect on the development of good working guardian dogs, complete with their regional variations. Every dog lives in two places: on wintering grounds in the warmer lowlands, and in cooler, mountain summer pastures. Lowland grasses dry up in the hot Mediterranean summer. The flocks *must* migrate to the fresh mountain meadows emerging from under winter snows. In the fall they *must* get out of there before being trapped by winter snows. If you don't move in time in the spring or in the fall, all the sheep die.

During the winter, Greek shepherd dogs rest on the plains at Thessalía, protecting their sheep from Greek wolves and jackals. Then, those very same dogs walk the three hundred miles through Macedonia to the Albanian-Kosovo border. Many of the shepherds are Muslims, and many speak Albanian and not a word of Greek, Macedonian, or Serbian.

I sit in conferences and listen to discussions about what color Greek shepherd dogs are supposed to be, or Turkish shepherd dogs, or Yugoslavian, or whichever. But I have seen wintertime Greek shepherd dogs transform into summertime Yugoslavian shepherd dogs, also known as beautiful Šarplanina mountain dogs, or beautiful Albanian mountain dogs. Today, with reemerging ancient countries with nationalistic spirits, the summer dog's heritage includes direct descent from the dogs of Alexander the Great's own flocks. During each season, the exact-same dog can mutate linguistically into three or four breeds, all in the same year.

The large livestock-guarding-dog "breeds" typically belong to these nomadic (transhumant) peoples. On every trip their dogs are capable of breeding with local dogs along the way. When they get to the high country they breed with shepherd dogs from other regions and other countries. Pups born on high pastures are sold and given to friends from other faraway places. A single dog's genes thus can be spread along the trail and then, through the puppies, be transported to regions the parent dogs have never visited. And all this can happen in a matter of months. In a single year, a single dog's genes can move thousands of miles. The Foundation for Transhumance and Nature in Switzerland estimates that there are over 77,000 miles of sheep trails worldwide, in width as much as 250 feet. Each migration averages from 370 to 620 miles, one way. Although some of the flocks are now transported by trucks—for example, in Spain, where the last big migrations took place in the 1950s—the animals are still moving great distances, twice a year, and spending long months in the new environment. And in the Middle East and Asia, sheep, and shepherds, often on horseback, still transport themselves.

The impact of the transhumance on dog genetics becomes crystal clear when you understand how many dogs are migrating over these long trails. On the Greece-Macedonia migration I watched, there were better than half a million sheep divided into morras of about three hundred and fifty sheep, a shepherd or two, and two to five dogs. Sometimes we would see two large, placid dogs walking near a shepherd, and behind them, a teenager towing two subadult dogs on strings. And sometimes we could hear baby pups protesting from inside a gunnysack dangling from a saddle.

Half a million sheep, then, would consist of fourteen hundred morras: at least fourteen hundred shepherds and as many as seven thousand adult guardian dogs.

The Macedonian transhumance may be average or small, compared with those farther east. If the proportion of shepherds, dogs, and sheep were the same as in Macedonia, then 414,000 shepherds and about

A little help from my friends. A shepherd boy on transhumance migration helps his dogs to complete the journey. Mutualism between people and dogs requires humans to help dogs to survive and reproduce and dogs to contribute their skills. Breeds of dogs originate when human help is directed nonrandomly at the surrounding dogs. Here the shepherd is aiding white dogs. Why he picked the white ones doesn't matter. In the next generation there will be a higher frequency of white dogs.

1,600,000 dogs accompanied the transhumance on their biannual migrations. Even farther east, China is the third-largest sheep producer in the world, with huge migrating herds around the Gobi Desert.

My guess is that well over a million adult sheepdogs are moving back and forth over three continents in a thousand-mile-wide band from the western Mediterranean to somewhere east of the Himalayas. Each year there is mixing and remixing of endless populations of dogs. Over the centuries, sheep cultures have repeatedly invaded new lands: Arabs into Albania; Sumerians into Hungary, perhaps all the way to Finland; Asiatics into Turkey. Each population gathers genes and brings them home to share with winter companions who have collected genes on their own in other locations. And they have done this twice a year for, what—four thousand years? They have been doing it for so long, the sheep have worn deep trails in the rock, as if glaciers had carved the passes.

For a population geneticist, the mental imagery of a million dogs migrating semiannually across Eurasia and Africa is simply mind-boggling. Every time someone starts talking about breed, my brain wobbles with the thought of the transhumant migrations.

THE TRANSHUMANCE:
EVOLVING THE SIZE AND SHAPE

Yes, I believe the ancestors of our modern livestock-guardian dogs arrived in each area with ancient nomads. I believe that the Macedonian dogs came with their Turkish Muslim shepherds years ago from Anatolia, and before that from Iraq and Iran and Afghanistan and China. And I believe that they came to places where there had been previous migrations, and also that they are migrating back to those places. Many of these migrations are not strictly discrete events; some may simply be disruptions of traditional journeys. Many of these shepherd cultures are much older than the nations they cross. Many of these shepherds speak languages different from those in the areas in which they travel or graze. In fact, many of these shepherds seem immune from international restrictions, and cross boundaries without passports as they have done for centuries. When a Masai warrior was asked where his passport was for crossing into Kenya, he said, "My earrings are my passport!"

Because these shepherd cultures are old traditions does not mean that their dogs have remained pure (i.e., sexually isolated) since the original migration. Actually, it means just the opposite. The importance of these long migrations is that the million dogs currently attached to the flocks are a homogenization of populations. The first mitochondrial DNA study I ever did showed that my border collies had identical haplotype patterns to the Šarplaninac I'd brought from Yugoslavia. In another study, an Italian Maremmano-Abruzzese I bought on the Gran Sasso had an identical haplotype to wolves from Romania and southern Russia. Having the same haplotype means that somewhere in the not-too-distant past, my border collies and the Šarplaninac had the same grandmother. In my Maremmano-Abruzzese, protecting my sheep in Massachusetts, was the identical haplotype of some great-great-grandmother whose haplotype also appeared in a wolf from Russia, probably descended from a wolf she met while on the transhumance.

The fact that these migrations are arduous means that the stresses of migration are factors in terms of natural selection. Like all other organisms, mountain dogs are selected by survival of the fittest, in the classic Darwinian sense. Any dog that doesn't or can't keep up with the migration is lost to the livestock-guarding-dog gene pool. In Italy, dumps along the way often have two or three dropout Maremmano-Abruzzeses skulking about. I still wonder why I spent good money for Maremmano-Abruzzeses in Italy when the dumps along migration paths are full of really fine-looking dogs that got left behind.

Dogs on migration have high mortality rates. On the Macedonian migration I saw many dead dogs along the way, mostly hit by cars. And the transhumant dogs come into contact with and spread diseases over great areas. Exploring Turkey for healthy dogs made me realize the extent of those puppy diseases throughout the dog world.

The dogs of the transhumance, therefore, have tremendous selective pressures put on them—by their activity, their feed, their breeding, and their mortality. One of the effects is on their size. They are typically bigger than the average village dog. Watching them, I can see why. First, bigger dogs have bigger strides. Bigger dogs cover more ground with each

Dog in the dump. This dog has dropped out of the transhumance, which, today, means that shepherds couldn't get it onto the truck. It is now lost and has little chance to reproduce in a sheep culture.

step. How many steps does it take to walk six hundred miles? It depends on how big the dog is. Each step contributes to wear and tear on the dog. Each step takes energy. To cover the distance with half the steps means a longer-lasting dog, a dog that has a better chance of making the round-trip. Such a dog is better adapted to keep up with the flock.

Second, their size enables them to endure a deficit of food; big dogs don't react to starvation as adversely as small dogs. Big dogs can carry more fat reserves and store more heat because they have lower surface-to-volume ratios. This is important, biologically. The bigger you are, the longer you can go without food. Animals that endure long, food-scarce Arctic winters (for example, the woolly mammoths and Irish elk of the glacial period) are gigantic species even within their taxonomic families. For mountain dogs, migration is like a winter. Human-waste resources become less stable. Each day the dump is left behind. On migration, people, dogs and sheep all have less food and less time to eat. It is better to start out with lots of reserve, and not take too many steps.

Third, big animals are more likely to survive than little ones the ravages of diseases, adverse weather, and accidents. Puppies, like children, don't have the body mass to survive rapid dehydration, and so they succumb more quickly to the effects of diarrhea. Big dogs are also better at tolerating low temperatures in those mountain pastures. Sudden storms with freezing rain can kill little animals fast. And it may be that big animals are less accident-prone. They cross streams and scale slopes at less risk, in part because bigger bones don't break as easily as little ones.

That is not to say big dogs don't have their own unique set of problems. Not the least of these is that big animals require more food. But in comparison with the wastes on Pemba, sheep cultures provide better-quality waste food. The by-products of sheep—milk, whey, dead sheep, sheep manure, and afterbirths—are rich in protein and fats, which support large animals better. And, like our Chake Chake curs, the dogs would be more protective of this high-quality food, sticking closely to this moving dump, and threatening approaching predators.

Really big dogs (one hundred pounds or more) have terrible problems getting rid of excess heat. Dogs like Saint Bernards and New-foundlands commonly suffer from heatstroke on a long walk. These two big breeds evolved in cold climates, where heat load is not such a problem. It is doubtful that such big dogs could survive a Mediterranean migration. On the first hot day, the hundred-plus-pounders would be popping blood vessels in their brains. Most of the dogs we saw on migration were in the fifty-to-eighty-pound range. Farther east, in semidesert countries, the flock dogs get smaller and sometimes have greyhoundlike shapes—which probably adapt them to getting rid of heat. Higher up, in mountainous countries such as Afghanistan and Nepal and into Mongolia, the medium-sized dogs have blockier shapes for conserving heat.

Again, it is natural selection in action, and each regional size and shape is a compromise between at least three factors: the length of stride that is most efficient, the amount of heat generated and stored, and the amount, quality, and distribution of the food source. The size and shape of flock-guarding dogs have to be adaptations to the particular transhumant niche they occupy.

It might seem intuitively obvious that bigger dogs are necessary to defend sheep against wolves, bears, or big cats. But that is only marginally true. Many shepherds around the world have small (thirty-pound) guardian dogs that appear to be just as effective. The twenty-five-pound Masai cattle dogs warn against lions, and the similar-sized Navajo dogs in Arizona are effective against coyotes and pumas. The dogs of the Damara in southern Angola and Namibia are indistinguishable from the local village dogs and protect against leopards, cheetahs, baboons, and a host of other big predators.

Very large dogs are not needed to protect against predators. The mythology about how the guardians defend livestock includes images of dogs out there fighting packs of wolves and being courageous heroes. I have no doubt that occasionally happens. But most of the time nothing happens, because predators don't like to go into a flock of sheep with five dogs watching the scene. Most of the time, guardians protect by being defensive, disruptive, and noisy.

It is rare that a predator will engage a dog in a fight. Actual physical engagement, with teeth, is not to a predator's advantage. Fighting is energetically very expensive. Neither predator nor dog has a lot of surplus energy to expend on dangerous activities. All animals have to assess the "cost" of fighting. The stakes have to be really high to risk it. Most cats—pumas, cheetahs, or leopards—will often retreat from even the smallest dogs. A wild predator injured in a fight—even one it wins—risks infection and an inability to hunt effectively, making it harder or maybe impossible to acquire its next meal.

Most predators are not behaviorally programmed to fight for un-caught prey. They will defend a carcass, but that is very different from fighting for the right to attack a prey.

Guardians warn stock and shepherd about the presence of a predator. The most important part of the equation is (and I can't say this loudly enough) that the predator is being warned. The predator is warned that it has been *discovered*. Someone is watching, someone is focused on this predatory behavior. As I pointed out in Part I, many species, including wolves, are shy about feeding if someone is watching. That one single fact in itself is enough to stop the hunting-stalking behavior of many preda-tors. Animals like cheetahs will often break off eye-stalk behavior if the potential prey discovers them sneaking up.

Stalking a prey, that is, sneaking up on it, while something is barking at you just doesn't make sense. In most cases, approaching and barking at a predator are enough to divert its attention away from the hunt. Most carnivores can't continue hunting if some dog is yapping at them. Further, a predator trying to focus on a fat lamb while the dog is trying to engage in tail-sniffing, territorial, or dominance displays cannot maintain the attack.

Shepherds, whether they have large or small dogs, will keep more than one dog, perhaps as many as five, for every 350 or so sheep. Wolves trying to penetrate the flock are faced with a cacophony of sounds com-ing from all directions. Many shepherds bell their sheep, which adds to the confusion. Bells are novel sounds for predators and can produce fear-avoidance reactions. The agitated sound of bells on the necks of sheep being stalked or chased also alerts the dogs and shepherds to a broken normal rhythm.

In the Old World, shepherds often tell me that guardians must earn their spiked collar by killing their first wolf. I asked a shepherd in Por-tugal about his seventy-pound dog sporting an ugly-looking iron collar, "Is he any good?" The shepherd replied, "He has killed many wolves." When I said, "But he doesn't have a single scar on his whole body," the shepherd responded proudly, "That just proves how good he is!"

It's a recurring theme in my life. I sit through lectures where testi-monials are given about how effective some dog of the national breed has been, taking on an entire wolf pack and singlehandedly saving the sheep. I sit, remembering my hours measuring the skulls of hundreds of wolves and livestock-guarding dogs and thinking how cute the latter's teeth are, and how small their heads are, and how weak their bites. Are dog brains so small that they can't see they are outclassed in a fight against a wolf? I don't think so. Every time I've seen a confrontation between a wolf and a dog, the dog looked scared.

I know dogs can fight with wolves. Blizzard, a Šarplaninac-Maremmano hybrid, fought a defensive battle against four wolves one night in Minnesota. The next day, I tracked the combatants for two miles with my heart in my mouth for fear they might have killed Blizzard. I found patches of dog hair and then wolf hair, giving evidence that sometimes Blizzard took a wolf down and sometimes they pinned him to the ground. And somehow he fought his way back to his feet. Blizzard finally got his back to a wall, a corner of a concrete abutment, and made a stand. And that is where I finally found him. The wolves had long since left.

He was fine. There was not a mark on him. Like many canine fights, the altercation had been mostly ritual. Dogs and wolves display their prowess to each other, but nobody wants to get hurt. I don't know what those four wolves were. Three of them might have been big pups that were not much of a match for Blizzard, the old veteran. They may have been with their old dad, who was feeling his arthritis that day. The image of the wild wolf on the prowl rarely includes its individual variations. In this fight, Blizzard figured out that he was outclassed, took up a defensive position, and he stayed right there. Even after the wolves took off, he stayed right there. He was in general a grumpy dog, but he wasn't a fool. He'd figured out that if these guys got serious he could become a pile of meat and fur. And each wolf had figured out that even though the four of them could take Blizzard, he might bite somebody seriously. Each wolf had to assess its own chances of survival, and the pups might not have had enough experience to be very sure of themselves yet.

The fight between Blizzard and the wolves was exciting. But that is not what I want to happen. Essentially, I don't want my dogs fighting. I certainly don't want them killing wolves or coyotes. The whole idea behind our project was to find a nonlethal method of predator control. The idea is not to kill or even displace wildlife. What's important is to keep the wildlife from killing sheep. My favorite guardians are the little yapping dogs of the Damara and the Navajo.

On my farm I have a little (fifty-pound) Maremmeno-Šarplaninac hybrid named Ellen who is the sweetest dog around. She wouldn't bite anybody, wouldn't fight anybody, and stays with our sheep. We have no losses or liabilities, even though there are coyotes in our forest and visitors in the barnyard. She is everything I want in a livestock-guarding dog. If I could walk to Washington, she could too. I tell people Ellen is a livestock guardian, and in my mind that means she has the perfect behavior and shape to do the job. This is because she grew up in my pasture with my sheep. She and I have a mutualesque relationship—I feed her, take care of her health, and give her breeding opportunities. And I think the reason I don't lose any sheep to predators is because she is there. Ellen is unregisterable by any breed club, but she is, to me, the result of thousands of years of natural selection, and the absolute best "breed" for the job.

BREED GENESIS: SELECTING FOR COLOR

Regionally, dogs often have distinct colors. The high frequency of a particular color can also be in part the product of culling. In an earlier chapter, I pointed out that color means nothing to dogs. Their reproductive games are played with their noses and behavior. What color the individual is plays little role in whether it breeds or not. There is an exception to this, which Cato points out. Good dogs should be well pigmented. Eyes, lips, and footpads should be of "good" color. In my experience, the blacker the better. Pink eyelids get sunburned, and pink footpads are softer and wear out more easily. When people insist on breeding white dogs to white dogs, they eventually create—at the very least—pigment problems. I always liked the Italian Maremmano-Abruzzese in my project except for that white coat. One of the best working Maremmano-Abruzzeses I ever had was named Anna, but because of her pink eyelids, she was constantly crippled with sunburn.

Some of the regional breeds (now) are predominantly white: Maremmano-Abruzzeses, komondors, kuvasz, Polish ovčarkas, Pyrenean mountain dogs. In most regions, white dogs exist, but at low frequencies. Cato preferred white. I suspect this is because white puppies are, for whatever reason, cuter in the minds of people than are darker ones. They are also rarer. In a heterozygous litter, on average, two pups out of eight will be white. I was looking at an Italian Maremmano-

Abruzzese litter one time and the shepherd said that the two white pups were the "pure" ones. He destroyed the rest.

It is fairly common for shepherds to cull litters to two pups. Livestock-guarding dogs tend to have big litters, which are difficult for a nomadic scavenger mother to care for. And it is difficult for a shepherd who is following the grass to attend to a favorite dog if she has a big litter. Culling also means that those two pups are more easily focused during the critical period on their mother and on the sheep rather than on their playful littermates. Hence those two pups will make better guardian dogs.

Culling of litters down to two is where the regionally distinctive breed colors are born. Shepherds have their favorite colors. Shepherds identify a sheepdog by its color. If I wanted to sell a dog or even advertise that I had good dogs, I would stick to the regionally favored color. Just as the Italian shepherd thought the two white pups were the pure ones, the Portuguese shepherd thinks the harlequin pups are pure, and the Turkish shepherd thinks the fawn with the black mask is pure. In each case the pups that survive the initial culling have the regional color pattern. If there were no pups in the litter with the regionally preferred color, then none of the pups might survive, or two pups of some other color would survive. It really depends on whether shepherds need replacement dogs or not. Often, shepherds will keep only the male pups. Males make better guardians simply and only because they don't have litters.

The outcome of culling litters down to two pups of the preferred color is that in a few generations, a high frequency of dogs will have coats of that color. Thus when a dog breeds with its neighbor, the neighbor probably has the same color coat.

When I was in Portugal in 1980, the Estrela mountain dog and the Castro Laboreiro were not very uniform in coat color, although jet black was common in the Castro. I couldn't find the other traditional breed, the Alentejo. Now, twenty years later, in response to political/ethnic events, the populations of the three breeds have grown and each has become distinct and easily identifiable by coat color. In the town of Castro Laboreiro in mountainous northern Portugal, most dogs now have a brindle pattern on a dark coat. But—the uniformity of the dog's conformation has not kept up with the uniformity of its color. The reason is simple: shepherds can identify the dog's adult coat color shortly after

Pure   Maremmano-Abruzzese. These are the pure Maremmano-Abruzzeses, I am told. It must be true because pure Maremmano-Abruzzeses are white, and the rest of the litter will be culled because they are not pure. The shepherd's definition of pure is not wrong, it is simply different than mine.

Pure Maremmano-Abruzzese. Around the corner is the rest of the litter, the not-so-pure Maremmano-Abruzzeses.

birth, but the size and conformation are impossible to predict in the neonate.

Recently, with the interest in preserving rare "breeds," the breed-identifying characters have become valuable for off-the-mountain sales. As the shepherds' dogs become valuable for other than their working ability, the shepherds pay more attention to what are perceived as breed characteristics and cull those puppies that don't have those traits.

In each region the high frequency of some color is accompanied by the development of a story about why that color is important. Like Cato reasoning about white, northern Portuguese shepherds believe the harlequin coat of the Castro Laboreiro makes the dogs invisible to the wolf because it blends in with the local rocks. The reasons for preferring white are many. Cato's lions aren't much of a factor in Italy these days, but white dogs, they say, blend in with the sheep, so the marauding wolves have to be careful. White dogs are preferred also because the shepherd can tell the dog from the wolf in the dark, and not club his dog hero by mistake. And some will insist that the original flock guardians were white, and white is pure and therefore a sign of antiquity. White signifies to many a state of noncorruption: white dogs aren't mongrels, is the intuition.

Well, the explanations make nice traditions. But closer inspection of the reality belies most of them. Nice white sheep are rare. Rather, their fleece tends to be dirty gray, picking up the color of the substrate. I can usually pick out a white dog among sheep from miles away. I have never met a shepherd who has accidentally clubbed his dog to death, or has even clubbed wolves to death in the dark. I have never met a wolf who has met a lion. I have never met a guarding dog with enough scars to attest to any necessity it may have had for fighting off wolves. Nor have I ever met a dog I thought could or would want to try to beat up a wolf (its small head, teeth, and brain must be intimidated by the idea of attacking big head, teeth, and brain). My best and smartest livestock guardians always showed modest, conservative behaviors around wolves and bears. Wolves and bears usually show reserved behaviors around a group of dogs. Who wants to get hurt so he can't forage tomorrow?

Most of the regional varieties are not white, of course. In the mountains, even the white Maremmano-Abruzzeses often have large yellowish or sometimes grayish spots on their ears or back. In most of the pastoral paintings I've seen in European art museums, a pure-white dog is rare. Color serves as a marker that the dogs that produced the puppies were shepherds' dogs. In some few areas the white pups survive the culling process. They also attract tourists, who purchase pups for very good money. Other regional color variations attract puppy buyers. Color is their best sign that the pup is an authentic member of the local shepherding livelihood. In some places, pups' ears are cut back, which is supposed to signify a true shepherd's dog. People with working dogs, however, know well that "a good dog can't be a bad color." The best Anatolian shepherd I ever bought was named Bernie. I bought him from a shepherd who knew dogs better than anybody else I'd met in Turkey. Although most of the shepherd dogs in that region were tan with dark muzzles, Bernie was colored like a Saint Bernard, which is a common color among village dogs.

WALKING HOUNDS

But of dogs there are two kinds, hunting dogs, which are used against wild beasts and game, and herd dogs, which are used by the shepherd.

—Cato the Elder, 150 B.C.

Hippolyta: I never heard so musical a discord, such sweet thunder.
Theseus: My hounds are bred out of the Spartan kind,
So flewed[1], so sanded[2]; and their heads are hung
With ears that sweep away the morning dew;
Crook-kneed, and dewlapped[3] like Thessalian bulls;
Slow in pursuit, but matched in mouth like bells . . .
—A Midsummer Night's Dream, Act IV, Scene 1

Besides the livestock-guarding dogs, Cato also wrote about hunting dogs 2,150 years ago. I call them the walking hounds. Beagles are a good modern example of a dog that accompanies a person into the field. In the

1. With hanging cheeks.
2. Sandy colored.
3. Flap of loose skin hanging from the throat.

course of the walk both person and dog end up chasing a rabbit or other quarry. Other walking hounds are the otter hound, mink hound, wolfhound, coonhound, and so on. Since there are over two hundred recognized breeds of hound, they may well constitute the largest number of distinct types of dogs. On Pemba I am a member of two genet-hound hunt clubs and have participated in a monkey-hound hunt club. (Genets are in the mongoose family, but superficially they look like big weasels.)

Walking hounds are essentially distinct from gun dogs (see Chapter 6), which are the pointers, setters, and retrievers. If you have a subscription to *Gun Dog* magazine, you won't see any mention of the walking hounds. Gun dogs have special arrangements of innate motor patterns. The walking hounds are more of a generic dog. Breed differences in behavior are primarily the result of the puppies' developmental environments. I think of them as generic dogs in the same way I think of livestock-guarding dogs as generic dogs raised in a sheep culture. Granted, the distinction between walking hounds and gun dogs is not always clear-cut. But the point is that regardless of selective pressures, a shepherd's dog or a walking hunter's hound will not express breed-specific behaviors unless strict attention has been paid to its developmental environment.

The walking hounds have identical evolutionary histories to the livestock-guarding dogs. They start as village dogs somewhere in the distant, or even in the not-too-distant past. As village dogs, they grow up in a hunter's environment. Those that "learn" to hunt well are favored by people. As favored individuals, they get better feed and better health care, which in turn may result in better reproductive opportunities. Those reproductive opportunities may frequently be with other local dogs that also are favored for their hunting behavior. In addition, hunters cull, or withdraw support from, dogs they don't like. Natural selection, human support, and culling produce regionally distinct dogs even though no one is actually breeding dogs.

The walking hounds are nestled here in this book between livestock-guarding dogs and sled dogs. They are here because their evolution is intermediate between the self-trained, self-selected livestock-guarding dogs and the sled dog, where every detail of the dog's morphology and behavior is scrutinized by the driver. The difference between the walking hound and the livestock dog is that the hunter consciously manipulates the puppies' environment so that the adult dogs not only learn how to hunt, but what to hunt.

Walking hounds might be the earliest example of useful dogs. By useful, I am implying the symbiotic sense of mutualism. These dogs could be the oldest in the sense that any village dog could be developed into a walking hound. Since the generic village dogs could have evolved several thousand years before sheep, the opportunity for dogs to be walking hounds preceded the opportunities to be livestock dogs. People could have raised (imprinted) and paid attention to the well-being of the dogs, and trained the generic dogs soon after the village dogs evolved. Since it is so easy to produce a walking hound from the generic village dog, Mesolithic peoples could have produced animals that were predisposed to hunt with them.

It is possible, but I doubt it happened until much later in the Neolithic age. I will come back to this question in a minute. But because people used village scavengers to hunt with is not evidence that they bred dogs for this purpose, or that there were breeds of such dogs in any sense sexually isolated from other dogs. Indeed, the great number of modern "breeds" suggests that it is a very easy leap from scavenger to hunter, and every region of the world and every era of history has repeated the process over and over.

In Chapter 2, I described village dogs that live in a commensal relationship with people. I used an anthropological technique in which I compared the life of modern Pembans with that of the hunter-gatherers of the Mesolithic age. I said that Pemban dogs were not strictly comparable to the original dogs, but I thought the selective processes and forces might have been illustrative of how it all happened. Although I said in that chapter that Pembans tend not to like dogs and avoid them, I was focusing on the hunter-gatherer portion of their population and ignoring (for the moment) a segment of the agriculturalist population that does like dogs (kind of), and uses them regularly for hunting. Pembans have walking hounds, and they use them to hunt vermin.

I was studying the feeding behavior of village dogs on Pemba when quite by accident I came across a monkey hunter with six dogs. They looked just like any other village dogs except they were following him, which is rare and unusual behavior for dogs on Pemba.

This man hunted green monkeys because they are considered a crop pest. The mango farmers had recently hired him to kill the monkeys depredating their groves. It was a job and it paid well. I had run into monkey-hunting dogs years ago on St. Kitts, in the West Indies. There the green monkeys ate sugarcane, and the hunters were paid a bounty for every one they killed. But on St. Kitts the carcasses were valuable as human food: "Monkey meat is sweet," I was told. When I asked the Pemban hunter about this, he just screwed up his face, as if I'd suggested eating rats.

Did he have good dogs? He told me that he had very good dogs and had caught many monkeys. He took his bare foot and rolled a dog upside down to show the scars on its lower jaw and neck. (He didn't really like touching dogs, and this footwork was an interesting compromise for displaying his pride in his dog's prowess as a monkey hound.)

I asked if this demonstration of scars proved this was a good dog. "No," he said, "I know I have good dogs because I always win the hunt club prize for the best dog." Well, you could have knocked me over with a feather. The hunt club prize?! I was amazed, because supposedly, as I described in Chapter 2, these people don't like dogs. And they don't. But at the same time, there are the dog-oriented hunters.

A good monkey hound. This is a good walking hound, not because he is scarred from many battles with monkeys, but rather because he is a frequent winner at the local hunt club. The owner has just rolled the dog over with his foot. He likes dogs, but he doesn't really want to touch one.

Hunters on Pemba hunt varmints and crop pests with dogs. Green monkeys are destructive to mangoes and vegetable crops. Indeed, monkeys can eat anything people can, and compete with people for food. The other major pest is the beautiful genet. It lives around the villages and steals chickens and ducks.

The important point here is that the dogs and the people are not hunting for food. Pembans are Muslim and have strict rules against eating monkeys or genets. In England, foxes and minks are the pests, and they aren't eaten. In America, we don't eat coyotes or rats but both are hunted by people with dogs. Hardly anybody eats raccoons. This is not unusual behavior for hunters.

In many parts of the world, farmers and ranchers hire hunters to rid their property of pests. In Portugal, wolf hunters parade their victims from town to town, and the shepherds tip them for killing such a terrible varmint. In America, ranchers will pay for hunters to come with their dogs, find coyotes, and kill them. In past decades the hunters not only got bounty payments from the ranchers, they could sell the furs. Terrier hunt clubs in the United Kingdom will go to people's farms and spend the day killing rats. In the old days, farmers would also pay for hunters to kill badgers, minks, and foxes. I interviewed a farmer in England who invited a lurcherman to get rid of the rabbits. Seven rabbits, he told me, eat as much grass as one sheep. The hunter got 450 rabbits, and the farmer claimed he had barely affected the population. Rabbits in this case are not considered game or food but simply a pest species. My favorite pest-species story involves the trout and salmon clubs that paid otter hunters to rid their streams of salmonid-eating "pests." Isaac Walton, the "compleat" angler, thought otters were the most terrible of vermin.

A lurcher, by the way, is a crossbreed between a greyhound and some other breed. The other breed is often a border collie, but can be almost anything—saluki, Bedlington, American pit bull terrier, Airedale. All are popular, and all have a special trait that blends well with the greyhound to accomplish a specific task. The reason I mention this is to reinforce the point that any dog will do if it has the ability to perform in a particular environment. The other point is that there is a long tradition among walking-hound hunters to create their own dog for specific purposes.

When people invite hunters to bring their dogs and rid an area of vermin, they do so believing that the hunters will be successful in alleviating their problem. But here is where our story takes a telling turn. In every location I have ever visited, every hunt is performed by people whose passion is not for the hunt or the kill, but for watching the dogs. The hunt is, or has become, sport. Like my monkey hunter who gets paid for hunting on a particular farm, the payment is just a bonus for what he would be doing anyway.

I joined two of the genet hunt clubs on Pemba, and not because I wanted to kill genets. I just love to watch the dogs work. I bought myself a really good dog, and I left her right there on Pemba. She is a top dog in a top club, and I'm proud of "my" dog. I send the huntmaster money to take care of my dog, and I hope she has puppies and I'll support them. I was such a good patron that they named one club after me (at least while I was there).

I love the hunt clubs on Pemba. I'll show up somewhat erratically and say, "Let's have a hunt." Everybody, including the village dogs, brightens up, arrangements are made, and a meet is on.

We collect at an appointed place. There will be twenty club members, a gallery of spectators, and fifty dogs, each following its own hunter. After some rallying ceremonies where tales of previous successful hunts are retold, off we go to where someone has spotted a genet. Then there is a lot of yelling and beating of bushes with sticks. Sooner or later a genet is awakened (they are nocturnal) and begins to move. A dog marks the event with a yipping cry, and the people and dogs are off on the chase. The excitement rises. There is a lot of shouting of instructions—head it off at the pass! Then we lose the genet. And all goes quiet. Then we find it again and dogs cry and people shout more directions. Sooner or later the genet is caught, maybe by a person who manages to hit it with a stick, or maybe the dogs bite it. A dog will pick up the genet and parade around with its colleagues tugging and pulling at it. Everybody rests (especially me) at the kill site and the huntsmen gather around and talk and laugh happily about the hunt, recounting moments of near misses, successes, and glory: "Did you see my dog when . . ." or, "I think such-and-such dog has lost it."

The genet is then impaled on a stick and triumphantly carried out of the forest. At some point the party stops again, a ritual fire is built, the genet is roasted and cut up, and each dog gets a tiny piece. (That is usually the signal for my annual contribution.) It is truly exciting, with good comradeship and wonderful dogs. That is why I am there. And it turns

The Ray Hunt. These are the best dogs in the world, not because they have caught the most genets but because ours is the best club. The best club always has the best dogs (if that makes any sense).

The Mink Hunt. Note the similarity of the English mink hunt to the African genet hunt.

out that is why everybody else is there—to watch the dogs and swap stories with your buddies.

Besides having a good time, is it useful work? Is it a symbiosis between people and dogs? Are the dogs chasing and killing genets performing symbiotically with humans for their mutual good, their mutual economy? Is the motive of the hunters to get rid of pests so there will be more chickens for starving children? Do I support the club dogs because they save chickens for this protein-poor country?

I don't think so. I've asked the question many times, "Why do you hunt genets [or monkeys]?" The first response is always "It is fun to watch the dogs work." I push and push for a better, more socially responsible answer. And I never get it. Then I rephrase the question: "Aren't genets a terrible pest?" And the light dawns as to what I'm looking for and they tell me what a terrible pest genets are and why they need to hunt them. Substitute "monkey" for "genet" on St. Kitts, and "mink" in Great Britain, and "coyote" in America, and it's the same story.

Fox hunters have a slightly different take on the scene. They will tell you that fox hunting is now tradition. They actually have to raise and release the varmint. One avid rabbit hunter in England is also the leader of the rabbit conservation movement. He doesn't want this terrible grassland pest to be eliminated, because there will be nothing left to hunt.

It seems to me that there are several lessons from our Pemba study of the walking hounds. First of all, the walking hounds are associated with the agricultural culture and not the hunters and gatherers. Could we interpret that to mean that the original use of walking hounds was a product of the Neolithic and not the hunting-gathering Mesolithic age? Could we contend that hunting dogs were from the beginning varmint and sporting dogs, and not used for hunting food?

Using dogs to capture food for humans is not unheard of. But I have always wondered whether, in the great scheme of things, meat-hunting dogs were ever very important. Although one popular hypothesis is that people originally selected dogs as hunting assistants, it never made much sense to me. Some claim that cooperation between wolves and humans, combining the power of the human brain with the senses and agility of the wolf, could increase the take-home pay for each—if they cooperated and shared the results. But I'm unaware that any of these concepts carry the calorie counts from start to finish, or even try to demonstrate an advantageous calorie intake per unit of effort for either species.

Mostly, people assume that since wolves are social hunters and people are social hunters, linking together would benefit both. They assume that tame, trained wolves would share a vital resource with humans.

From a biological viewpoint, when I calculate the energy costs to humans and dogs cooperating in food-gathering behaviors, I don't find a significant advantage for either. A formal study needs to be done that would measure how much energy a dog puts into a hunt, and add how much energy the human puts into it, and then divide that figure into the number of calories returned by the captured-prey species. The formula might be: # of calories in a rabbit ÷ # of calories expended by humans and dogs to capture rabbit.

If the answer is greater than one, then the hunting mutualism would be confirmed. In my own dog-hunting experiences, five beagles and me chasing after a rabbit need caloric subsidies to sustain this activity.

With this in mind, when I contemplate the evolution of highly specialized retrievers from some wolf ancestor, I discard the wolf-into-hunting-dog model. I assume that in the late Mesolithic period, both people and wolves didn't have the sumptuous food-reserve budgets that my retrievers and I have. For the ancients, the energy in–energy out formula did not include the luxury of heading for the grocery store if the game got away. Besides, where did they get the energy to invest in this taming-of-the-wolf experiment? Wolves of the late Mesolithic, like any top-level predators, were on a very tight energy budget. After all, twelve thousand years ago there was a noticeable decrease in the size of wolves, suggesting that the environment had become less bountiful. If most wolves are or were starving to death (which is the basis of the Darwinian assumption), then sharing food with another species is not going to be a top priority. I expect that Mesolithic hunter-gatherer populations had to pay tribute to the Malthusian theorem, that increase in human numbers outstrips food production or availability. The average size of humans also decreased twelve thousand years ago, further suggesting a decline in the amount or quality of food. The human population must have been limited by food competition. My guess is that hunter-gatherers behaved in energy-conservative ways, just like other animal species. Some fellow experimenting with a caveful of hardly socialized pet wolves doesn't sound energy conservative to me.

The history of hunting with dogs, especially hunting large animals like deer, where the energetic payback might be better, shows this activity to be the privilege of the sporting wealthy. It is the activity of people who have access to surplus resources. It takes a lot of work and time to train a dog, and it takes knowledge to breed a dog that can do the job. And it doesn't always work that the dog turns out to be a successful dog. I just don't believe that dogs were ever significant as mutual symbionts in food gathering, but I shall remain open-minded.

But let's ask the question whether the original hunting dogs could have been varmint dogs. The more I observe the walking hounds, the more I convince myself that the original walking hounds were varmint dogs. If that is true, then it places mutualism with these dogs firmly in the Neolithic period (less than ten thousand years ago) rather than the Mesolithic. Varmints are animals that steal our domesticated crops. Once domestic crops were planted and became a significant portion of the human diet, varmint hunting made sense. Pests are ever-present around a village. They are a nuisance and they compete cunningly for human food.

Does varmint hunting satisfy the energetic requirements of the participants? Does it pay village people to keep a pack of varmint dogs around? In other words, is the answer to the formula—pounds of food saved by killing varmints ÷ energy required by humans and dogs to find and kill varmints—greater than one?

Maybe—but again, I doubt it. I realize that one of the main chores (energy expensive) of any farmer is protecting the crop. Western ranchers invest heavily in killing coyotes and protecting sheep from a variety of predators. One of my friends spends better than 50 percent of his ranching time protecting sheep from varmints. But, figuring that on Pemba the energy expended by the fifty dogs and the thirty men running around excitedly trying to catch a four-pound genet is in some way equal to the amount of energy contained in all the chickens that genet would have eaten in a year, is a tricky calculation. I suppose it is possible, though—after all, these Pemban dogs are still little dogs and still primarily living from village wastes. If, however, the English mink hunters are buying dog food and transporting dogs by trucks over hundreds of miles, then there is no way to balance the equation. The honest answer is that hunting with dogs is simply good sport.

The first lesson from the modern Pemban hunter-gatherer societies is that it is not their hunter-gatherer culture that supports these hunting-dog activities. It is their agricultural activities that initiate the evolving mutual symbiosis. These hunting dogs are not supplementing the food-gathering activities of humans. The mutualistic relationship between people and dogs is centered around crop protection. One could argue that since dogs are the beneficiaries of the wastes from the crops, participating with humans is an example of true mutualism.

The second lesson I learned from participating in the Pemba hunt is that hunting is initiated by people. For the most part, dogs show little natural interest in hunting. In order to be a hunting dog, wee puppies have to be removed from their natural environment and raised during the critical period in a special environment that predisposes them to hunt as adults.

Even then dogs rarely go, wolflike, searching for game. They lie around the village eating waste products (some of which their human hunters save for them). The hunt commences with the humans rallying at some predetermined location with their dogs and then proceeding to the hunting area. The dogs are following the people. It is the people hunting excitedly that motivates the dogs to hunt. It is *social facilitation.* The dogs are hunting because the people and other dogs are hunting. They are hunting genets because people and other dogs are hunting genets. They are killing genets because people and other dogs are killing genets. If my Pemba hunt friends walked to a potential genet territory and yelled and screamed and startled a genet, but then turned and headed for home before the genet was caught, the dogs would follow the people home. Some of the modern walking-hound breeds will hunt on their own. In these cases, the motivation to hunt is internal rather than external. They become more like the gun dogs in Chapter 6.

Walking hounds are not—repeat, not—hunting to eat. Dogs don't participate because they might get fed at the end. And, indeed, when they catch and kill the genet, they don't eat it. There is no feeding frenzy. They don't seem to have the dissect motor patterns. If the people cut open the genet, the dogs will eat it. Again, they seem very like livestock-guarding dogs in this respect. And for the sled dogs in the next chapter, the immediate reward for the dogs is playing with other dogs and people—playing in the sense that there is no biological reward in terms of food or reproduction from participating in this energy-expensive behavior. It is sport for the dogs as well as the people. They hunt to fulfill social needs.

Are these Pemba dogs different from our modern, carefully bred

Western hounds? I don't think so. I think they are just as sophisticated. With a television film crew, I participated in a British mink hunt and a Pemba genet hunt. The British let the dogs out of the van, walked to the stream, blew horns, and made strange noises. The dogs got excited and followed the people. There were only two differences of any note between the two hunts. First, the Pembans didn't dress up in red and yellow jackets, and second, the Pemba dogs caught their mongoose.

The third lesson from studying Pemban hunter-gatherer-agricul-turalists was the revelation of how one develops a species-specific genet hound. How can you get a dog to chase only genets, or foxes, or minks? Here again, I doubt there is any difference between the ancient and modern systems. Start with any village dog growing up among chickens and cows. Start with tiny puppies. Tie them up and feed them. Tease them with a dead-genet skin. As they grow up, take them on the hunt with other dogs. Encourage them to hunt. They will learn how to hunt and what to hunt from other dogs. From then on, each time the dog goes out, the hunt will reinforce the focus on the specific quarry. This script sounds a lot like the Navajo recipe for raising a livestock-guarding dog.

When I went on the hunt in Pemba with the fifty dogs, not all of them were hunting and it seems to me that at the end some of the dogs had already gone home. When I interviewed the mink hunters in England I got the same story. Mink hunting in England has recently become popular. Minks were introduced to England only a few years ago. At the same time, otter hunting has been discontinued because otter populations are dangerously low. The newly formed mink-hunting clubs have drafted dogs from the other clubs, dogs such as foxhounds, otter hounds, and rabbit hounds, and have "taught" them to hunt mink. "We don't have a mink hound yet," says one expert. "It sometimes takes a dog years before he gets the mink message," says another.

To call them a mink hound or genet hound suggests breed: "We don't have a mink hound [breed?] yet." But maybe the respondent did not mean breed in the AKC sense. The fact that some hunts are composed of dogs with uniform conformation, color, and behavior does not mean they are a sexually isolated population of dogs especially adapted to hunting mink. In a really fine hunt, all the dogs will be the same size, have the same cry, be of the same color pattern. This is important in the best hunts because if all the dogs are the same size and shape, they will run at the same pace and with the same motions. That increases the likelihood that they will run together. That in itself increases the mob effect and the self-stimulation to carry on. If some dog gets too far ahead or behind because it is a faster or slower runner, it loses interest and goes home. If all the dogs in the pack are homogeneous, with like behavior, then the excitement is maximized.

The first step in hunting-breed formation is to develop a hunt. Within the framework, some hounds will be better than others. Some dogs will follow the huntmaster, or find the mink, or—of great value—mark the mink better than others. Some will do this because they work together with other dogs. The dogs that stand out will get better care, better training, and ultimately more opportunities to reproduce. Those which stay in the village or don't mark the mink might be culled, or at least not be included in the breeding program. Like the transhumant livestock-guarding dogs, some of these hounds will simply get lost or abandoned.

Each new generation of hunting dog is better, and also looks more and more alike. It is very much the same process of breed development as for the sheepdog. But note, these carefully selected dogs are not a breed. Individual dogs do not have intrinsic (genetic) capabilities beyond their own hunt. If I took a really good dog and put it into a different hunt with dogs of different size, it might be a complete flop, simply because it can't work uniformly with these other dogs.

Getting a dog to focus on a single quarry species is the result of two separate processes. First, the pups grow up in the village or on a farm and they are imprinted on the farm animals. Modern foxhounds, for example, are raised by "puppy walkers." This is often a local farmer who takes care of the pups right on the farm, through their critical period of social development. Thus the dogs cannot and will not show predatory behaviors to farm animals, just as livestock-guarding dogs don't hunt sheep. The resulting adult foxhounds can chase the fox right through a farmyard without swerving toward the fleeing domestic stock.

Second, the dogs learn to hunt the target species. From an early age they are taught by various methods, including the dragging of a carcass lure. They learn what the fuss is about and which species is to be tracked. Primarily, however, they learn through social facilitation with other dogs. It is easy to train a sled dog if you have a team. It is easy to train a mink dog if you have a mink hunt. Once the British develop a good mink hunt, specialized mink-hunting hounds will quickly follow.

The hunt takes on a conformation of its own. The local hunt becomes a tradition that involves the entire community. It is a cultural habit. Repetition and anticipation generate excitement in dogs. In the next chapter, I'll describe a similar social facilitation existing within sled dogs, which run without chasing anything. These evolving mutual relationships with people of walking hounds and sled dogs are not based on modified wolf-pack behavior, but simply on self-stimulating social relationships between the dogs and the people. Food is not the goal; the activity itself is intrinsically rewarding. It is good sport.

In many hunt clubs one person cares for and develops the dogs, while others (like me) buy in. The pride is in keeping the dogs in a very good pack, one with uniformity and identifiable features. A unique color becomes the symbol of a particular hunt club, a particular expert, or a particular region. One can find wonderful examples of these color markers among the foxhounds, and even in the American coonhounds.

The idea that the marker color pattern is indicative of an intrinsic quality of a dog is a major misconception. It isn't an American foxhound that is being marked, it is a hunt club, and the dog is only as good as his club. Many modern breeders select dogs, in prezygotic matings, as if the coat color were symbolic of a good dog. The only good thing about the dog is that it is sixteen inches tall, like all the other dogs in the hunt. My friends who hunt with dogs for mink and genets are never breeding for color, but rather choosing good dogs that have that color.

It was the same with the mountain shepherds who did not cull the puppies from good females that had the preferred color. Similarly, Pemban huntsmen not only give good dogs more care, but they give the puppies of good dogs more care. These puppies have value. Whether hounds or shepherd dogs, puppies from notable dogs can be easily sold. In many of these rural, agricultural societies, the descent over the generations is matrilineal. Usually, nobody knows who the father(s) are. In African hunt societies, there is recognition of good mothers or even a good pack. A well-bred puppy can be from a good pack with no recognition of who its parents are. After all, if hunting is learned, then better hunting is learned from better dogs.

Selecting puppies from good mothers is a form of breeding. It isn't exactly prezygotic selection, but it is getting there. By having mostly hunting dogs in the area, the best female is most likely to get bred by the best hunting dogs. Again, the parallel with shepherd-dog husbandry is striking. Going back to Cato's advice to his shepherds on how to buy a good flock guardian, we read, Don't buy one from a hunter. And, if you are a hunter, you shouldn't buy from a shepherd. Once the pup is imprinted, you can't get it to behave another way.

What if we calculated the mutual value to each species (hounds and humans), and determined that there was no ecological benefit to hunting varmints? It turns out to be just plain sport. The vast number of dog people since the beginning of dogs just like to see the dogs run. The benefit to the dogs, then, is that they get better care and become valuable. The benefit for people is psychological—and not just for the hunter but for the community. The farmer feels cared for since the hunters and dogs are trying to rid the area of pests. The hunters are local boys showing their prowess as hunters and dog people. They take pride in their hunt and everybody else likes to watch. During the filming of the mink hunt in England, a hundred hunt-related people spent the day running around in the rain.

If I were going to contend (which I do) that pet and companion dogs bestow psychological benefits on their human companions, wouldn't I have to extend that consideration to all the people who take pride in the local hunt? I have a good dog and I feel better about myself even though she is in Africa. I have a good dog and other people want me to belong to their club. I like to watch dogs run, which connects me to communities of people that also like dogs. Certainly the zeal hunters show for the activity suggests it is very important to the participants.

What is the benefit for the walking hounds to be a part of this symbiosis? Is it true mutualism for the dog? I think that because the walking hounds are supported and cared for and valued, they probably have a better quality of life. To be born of a good mother probably increases the chances of survival, and perhaps longevity. This would be especially true on an island like Pemba, where the commensal scavenger isn't liked. Although dogs are spread out across a village fairly thinly, a hunter's yard may be full of dogs. In other words, the dogs are nonrandomly distributed across Pemba. The hunting dogs exist in higher concentrations than could be supported by only village garbage. Hunting dogs benefit from their talents.

The important message in this chapter is that any dog will do for any job if raised and trained properly. Brian Plummer in Scotland has a pod of King Charles spaniels that tears across the countryside searching for bunnies and kills them. Here is a hunt that is seriously uniform, to say the least. These spaniels have been a lapdog for centuries. They have not been selected to be hunting dogs. But Brian, who prides himself on being a good dog trainer, uses them to illustrate his point: if you think about it and work at it, and you know your dogs, you can teach any breed to do any task.

I made the case to Brian that any dog can do anything *only* if it is socialized correctly when a tiny puppy. I was sure Brian would see this as an exception to the rule that you can train any dog to do any task. He didn't seem to. Then I asked Brian if one could train King Charles spaniels to hunt lions. Or could King Charles spaniels join a transhumant migration, and guard sheep against wolves, even if they were socialized properly? Well, Brian's a good dog man; he knows dogs. He just gave me a big grin and asked, "What do you think?"

Brian Plummer and King Charles spaniels. Much of a dog's behavior is shaped during early development. These rabbit-killing spaniels are Brian Plummer's testament that you can train any dog to do anything.

CHAPTER 5

The Physical Conformation
of a Breed

SLED DOGS—HOW DO THEY RUN?

There are hundreds of breeds of dogs. Each performs some task. Each of them has a unique conformation. In theory, the unique shape allows the breed to perform its particular task better than any other breed or animal species. Another way of saying this is, continual selection of superior performing animals results in a unique shape.

When the distance is over ten miles, modern racing sled dogs are the fastest animals in the world. Nothing can outrun them for ten, a hundred, or a thousand miles. Many animals might be faster for a couple of yards, a hundred meters, or a quarter of a mile, but once the distance gets serious, racing sled dogs are untouchable.

Cheetahs, antelopes, and wildebeests all have faster speed records listed in the almanac (70 miles per hour for the cheetah, 61 for the antelope, 50 for the wildebeest), but these measurements were made over a quarter of a mile. They cannot sustain these speeds over much longer distances. Some cannot run even a mile at top speed.

In the animal world, humans are very good long-distance runners. Twenty-six-mile marathons are reasonably easy, and thousands of people enter such races every year. The current record is about two hours and six minutes, which is 12.5 miles per hour, or 4.8-minute miles. The horse can do better, although not many horse races are as long as twenty-six miles. The Qatar (Arabia) International Desert Marathon is run over the standard marathon distance (26 miles and 285 yards, or 42.2 kilometers). In the 1997 race, the winner, an Arabian, won in one hour, twenty-four minutes, and one second—a time of 18.7 miles per hour, or 3.2-minute miles.

In order to win most sled dog races, the best teams have to run 3.2-minute miles for near-marathon distances. That's on snow. That's uphill and downhill, through the woods, across streams, and around sharp curves, all the while pulling two hundred pounds of person and sled. In ideal conditions—on a flat, straight, packed trail—there is no reason a team could not run 2.5-minute miles. That would be twenty-four miles an hour, almost twice as fast as a human.

And then, unbelievable for a human or equine runner, the dog team runs the twenty- or twenty-five-mile trail again the next day. And sometimes still one more day! The World Championship Sled Dog Race held in Anchorage, Alaska, is twenty miles per day for the first and second days, and twenty-seven miles for the third day. For the winning dogs it isn't exactly a piece of cake, but what other mammal can equal this performance?

Then there is the Iditarod Trail Race in Alaska, an annual commemorative race of eleven hundred miles. When it was first run, in 1973, there was no trail. The dog drivers began in Anchorage and headed for Nome, with several checkpoints along the way. The drivers wore snowshoes, and had to pack down a snow-choked path ahead of the dogs, cross rivers, and watch out for moose. The winner that first year took twenty days, forty-nine minutes, and forty-one seconds. Now, nearly three decades later, the trail is packed by snowmobiles, has passing lanes and bridges. It is a lot safer for drivers and dogs, and a lot faster. The distance is still the same, but the winning time (and current record) in 2000 was nine days, fifty-eight minutes, and six seconds. That means teams average well over 125 miles a day, for just over nine days. That is like running five marathons a day for nine days! One twenty-four-hour rest and two eight-hour rests are mandatory. What other mammal could even imagine this feat? My job here is to describe how a dog can do that.

Modern sled dog racing is a pinnacle of mutual achievement between dogs and people. Symbiotic mutualism portrays two species living together for the mutual benefit of both. Dogs perform some task that humans cannot do. Humans aid and direct the dogs' activity, which accomplishes a goal for the person. For the dog, the reward is the social interaction with other dogs and people, as well as the enhanced opportunities for feeding, reproducing, and staying out of trouble. The good ones become part of an evolutionary process—they evolve the perfect shape. Such a mutual relationship is said to be functional. The dog's activities coordinate with human activity, culminating in advantages for both.

In biological terms, true mutualism denotes that both species adapt (change shape) physically or behaviorally. They change form genetically in order to live together efficiently. It seems, however, that mutualism between dogs and people requires dogs to change genetically, both in form and behavior, while humans maintain their form and learn to modify their behavior. Humans can back out of the relationship with no loss of genetic fitness, but the specialized dog is stuck with that new shape. In short, except in rare instances, if all dogs were to die, human life is not threatened, whereas if all humans died, the domestic dog could not survive in its present form. This relationship, then, gives a twist to traditional mutualism. In fact, it is not really mutualism in the strictest ecological sense, because one of the symbionts has not changed genetically in order to live with the other. Our relationship with the domestic animals is not what a biologist would define as coevolution, where each species is modified by the process. Sled dogs are genetically adapted for pulling a sled while the human only has to learn to ride it.

In order for sled dogs to have achieved their status in this specific niche, they have evolved from the basic village-dog shape to a highly specialized, proficient shape. At first, for sled dog drivers, it was "any dog'll do." Then, as the speed and endurance of sled-pulling dogs became more important, the dogs started to differentiate themselves into specialists. Of course, they didn't do this consciously, nor, perhaps, did the drivers. But then, as the generations passed, within this population certain dogs began to stand out that pulled faster or longer than other dogs. The contemporary result is perhaps artificial selection at its best. Today's sled dogs resulted from dogs captured or adopted from diverse niches; the originals were most commonly thought to be "northern" breeds, such as Siberian huskies, malamutes, or Greenland huskies. But the truth is, there is a good sprinkle of other breeds in there, such as retrievers, setters, or greyhounds. The unique shape of the modern sled dog was rapidly achieved by crossbreeding.

What makes the efficient shape absolutely fascinating to me is that many of the important, special features of sled dogs are unknown to their users. Drivers "intentionally" selected for superior working animals, but they selected on the basis of performance, without knowing exactly what physical characteristics support the abilities to pull a sled fast.

In the Alaskan gold rush of the early 1900s, thousands of people traveled great distances, in an inhospitable landscape, carrying their gear and sometimes themselves on sleds pulled by dogs. Quite simply, a person's survival depended on the dogs hitched in front of the sled. Dog drivers hauled freight, tended trap lines, delivered mail, and policed a vast area—making their living by driving dogs. The individual dog puncher became proud of his abilities to manage a big team. Although the ultimate test was survival, there were sporting competitions also. During the early decades of the twentieth century, at "fur rendezvous" celebrations, trappers sold their furs, socialized, and competed with their dogs for prize money and professional prestige.

Sled dog racing quickly became important as a sporting event. The world changed rapidly, and machines replaced the dog team with faster, more efficient winter transportation. But the sport races were still fun, and they became an economy unto themselves. A top driver could make a year's salary in prize money in a single event. A superior dog took on great value and could be sold to would-be competitors for big money. Owners of the best dogs could win the big prizes, and then put these dogs up for breeding and sell their puppies. Dog drivers were paying close attention to how their dogs behaved. The best dogs were being bred to the best.

Because of this, sled dogs are faster today than they ever were. Racing times have steadily improved. The winning dogs of twenty years ago would hardly place in a race now. When I started racing in the mid-1960s, a four-minute mile was good enough; but when I stopped driving fifteen years later, it took 3.2-minute miles to win most races. The drivers of today know so much more about dogs and training than did the great drivers of the past that they are consummate professionals. None of the pros is nostalgic enough about the past to wonder where the great dogs of the Eskimo have gone, or to think they should go to Siberia to find some hidden supply of fantastic dogs.

In fact, the teams that are winning the top races today are not Siberian huskies, but the so-called Alaskan husky, a hybrid of whatever runs the fastest. They tend to be long-legged, racy-looking dogs. They often have the pricked ears and coat coloring of a husky, but those are superficial traits that don't have much to do with running fast. It is their gait that pinpoints their success. The Alaskan husky's front feet barely leave the surface of the snow, skimming forward to grab the next stride. It looks almost possible to balance a tray of crystal on their backs, so level and smooth is their motion at top speed.

Crucial to top speed over long distances is an economy of motion. That economy is the product of the dog's conformation. No dog will make the team without the proper conformation. When I described the Pemba hounds, in Chapter 2, I used twenty-two to thirty pounds as my "breed standard," simply because that is what they are in nature. Were I not a biologist, I might not realize that the uniformity of the dogs is due to their adaptation to a village-dog economy. The standard for the Siberian husky states that males should not exceed sixty pounds. I assume that is because dog drivers have found by trial and error that animals over sixty pounds cannot do the job as well as the lighter ones.

I explored in Chapters 2 and 3 why twenty-five pounds is an adaptation to village-dog economy, so now I'll look at why less than sixty pounds is adaptive to a running sled dog economy. It is not intuitively obvious why a bigger and stronger dog would not work better. To tell the truth, I could never find anybody who could explain it to me. "Big dogs don't have any endurance," the pros would tell me. So, why is that? As a biologist who wanted to win sled dog races, the subject of size and shape fascinated me, so I studied breed conformation in detail.

The dog family in general contains phenomenal running animals. They are cursorial (specialized for running) digitigrades (adapted to run on their toes). Wolves, coyotes, and jackals move easily over the countryside. Most of them have a trotting gait, like a bird dog's, and explore vast amounts of territory, noses constantly sniffing for prey.

Just to illustrate how easy and natural it is for a sled dog to run, take the case of Joanie, a husky–border collie crossbreed. Joanie was a "walk on the leash once a day for exercise" city dog who was spending a weekend with us. On our rural acres she ran loose and did what she wanted. This was during the years of our big sled dog teams, when we had frequent training runs. The first day of her visit, Joanie followed one of the teams out of the yard, jogging along behind it for fifteen miles. We then took twenty minutes to unharness the team, put a new team together, and head out again, with Joanie jogging along behind. But alas, she couldn't keep up this time. She quit after five miles and went home. She didn't seem any the worse for what was probably her first twenty-five-mile jog. Try to imagine any other unconditioned animal in the world going on an odyssey of that length, doing four-minute miles, out of the blue. A human would be in the hospital; a horse would be dead. Neither would be up the next day, but Joanie was frolicking around, ready to go again.

One of the most important elements of an animal's ability to sustain high-intensity effort for more than a few minutes is its maximum oxygen consumption. Getting oxygen to working muscles, and using it, is crucial for the metabolism that produces the energy. The fact that dogs are a red-muscled animal simply means they have many, many blood vessels, giving them a large capacity to deliver oxygenated red blood cells and food to their muscles. Contrastingly, cats are a white-muscled animal. They don't have very many blood vessels. Instead, they have many more nerves enervating each muscle, making a cat's muscles look creamy-white compared with the dog's darkly red muscle.

The cat moves more quickly, and makes very fine and precise motions—but not for very long. Imagine a cheetah running full tilt at 70 miles per hour and, without spoiling its stride, reaching out with a front paw and banging a gazelle on the side of the head. The eyes and everything else about a cat's nervous system are proportionately larger—much more phenomenal compared with a dog's. But the way to catch a cheetah, the fastest mammal in the world, is just to run it down with a dog or a horse. In the old days a horseman would chase a cheetah until its muscles ran out of oxygen and energy, and the poor beast could only lie there panting as the human bagged it—and off to the zoo. The dog, with its thousands of blood vessels, can deliver nutrients to all parts of the body quickly and continuously. Sugar or fat delivered to and stored in the muscle is burned—combined with oxygen—creating the energy needed for continuous muscle contraction.

Oxygen is carried in the bloodstream by hemoglobin, the red blood cells. Hemoglobin, when filled with oxygen, is bright red. When we say a dog is a red-muscled animal, it's because we see the oxygenated hemoglobin in the blood vessels of its muscles. A good running greyhound or sled dog has so much hemoglobin that its blood is thick and sticky compared with that of other species.

One author, writing about greyhound hemoglobin, marveled that it was possible for the dog's heart to pump such sludge. A human athlete with that much hemoglobin would immediately be suspected of blood-packing. This is a process of withdrawing a pint of an athlete's blood long enough in advance of a race for the athlete's system to replace the blood. Then, a day or two before the race, the extracted pint of blood is put back, giving him 10 percent more hemoglobin. It is said that this technique has been responsible for a few Olympic victories. Dogs actually have a system for storing hemoglobin, which can be released as needed.

Another trait of sled dogs that makes them the best in a sled dog race, faster than a greyhound or any other breed, is their size:weight ratio. Although greyhounds speed down the oval racing track at thirty-six miles an hour, they are sprinting for a very short distance—tracks are between five-sixteenths of a mile (550 yards, or one-third mile) and seven-sixteenths of a mile. There I would be at the back of my sled, slogging through mushy snow in the pouring rain, pushing the sled and urging on the team, and after the race someone would say, "Why are you running those puny little fifty-pound mongrels? Why don't you get big, strong dogs like malamutes? Why don't you use Saint Bernards?" Or, when a driver would comment that the dogs had been slow that day, one would hear, "Why not use greyhounds?"

At the simplest level of explanation, the answer is that sled dogs have the perfect size for the job, and greyhounds, malamutes, and Saint Bernards don't. Sled dog conformation means no wasted motion while pulling a sled. The implication is that greyhounds, malamutes, and Saint Bernards have shapes that would waste motion when pulling a sled fast. The economy of motion enables the sled dog to run almost effortlessly while pulling a sled at a good speed for an entire long-distance race. Or, one might say, it allows them to perform this task more efficiently than any other breed. Since they run so efficiently in this environment, they can last longer than an animal that has excess motion. Greyhounds, malamutes, and Siberians, for different reasons, are awkward at pulling a sled at the best racing speeds.

If someone showed up with a team of beautifully conditioned and trained dachshunds, people would smile because they know the dachshunds can't win. I would smile if someone showed up with a team of malamutes, because I'd know they didn't stand a chance. I'd almost rather take a chance with the dachshunds. I say "take a chance" because it would be cruel to race dachshunds or malamutes against Alaskan huskies.

Neither of those breeds is built for that kind of exertion over those distances. Not only is it not possible for them to run at Alaskan racing speeds and distances, but they could be badly hurt in the attempt.

Running while pulling a sled is a complicated behavior. Running fast is not learned, it is genetic. A dog can't *learn* to run at racing speeds; it has to be *able* to run at racing speeds, and then conditioned to do it. If the dog is able to run at racing speeds, then it can learn to do it. If a dog doesn't have the correct running shape, it cannot be taught to run. I can't teach a dachshund to run even one 3.2-minute mile. For a different reason, I can't teach a malamute to run long distances. If I conclude, for example, that the dachshund's legs are too short, then I get a long-legged dog and solve *that* problem. Greyhounds have really long legs; why aren't they just a long-legged, long-bodied, really fast dachshund?

Which takes us back to: Why don't you get a greyhound to pull your sled fast?

There are two essential reasons why greyhounds cannot win sled dog races. One, they have an inappropriate gait for pulling a sled. And, two, they are too big and heavy.

Let's look at gait first. Greyhounds run with a double flight. (A flight is when the dog has all four feet off the ground during its forward leap.) A greyhound "flies" off its back feet, lands momentarily on its front feet, then cantilevers off its front feet into the air again while reaching forward with its back feet; it lands on its hind feet and springs forward again into the air. A human runner does the same thing while running, leaping from point to point, spending time in the air with no contact with the turf.

The double flight is very fast—and very unstable. Greyhounds fall down a lot. A fall at thirty-six miles an hour means at least a race lost, at most a permanent injury. The reason English coursing hounds—called lurchers—are crossbreeds of greyhounds is because pure greyhounds are susceptible to injury. In contrast to a greyhound, a racehorse has a single flight. The horse cantilevers off its front legs, flies through the air (very briefly, almost imperceptibly), and lands on its hind legs. But now it stretches forward and puts its front feet down while keeping its hind feet in place. For most of its stride, some or all of its feet are on the ground.

The reason the greyhound and the horse run in their particular styles is due to their bone structure, musculature, weight, and training for a specific task. The horse cannot jump onto its front legs at full speed. All that weight landing on those spindly, unbending, shatter-prone front legs is at the limit of what bone, muscle, and blood can stand. A cat or dog is suspended between the shoulder blades and the backbone in a loose-fitting sling of muscles, giving it a shock absorber. But the weight of the horse requires a much tighter musculature. It is the difference between the suspension of a luxury car and a dump truck.

Technically, what the horse does is part-walk and part-run at some point in the stride. In contrast, the double-flight gait of the greyhound is a true run. In a human walking race, one foot must always be on the ground, which means that for some brief instant both feet must be on the ground. There must be no flight, no leaping forward in a walking race.

Walking and running are different gaits and are based on different principles of physics. Walking is a falling motion, with gravity providing the acceleration. The individual leans forward and starts to fall down, and then moves a leg forward to stop the falling. This leg raises the body back up to the original height. In a true walk the leg doesn't provide any acceleration forward. The speed of a walk depends on the acceleration of the body by gravity (thirty-two feet per second squared) and the length of the stride. How far the animal or human falls forward on each stride is a large part of the speed it develops.

Running is leaping forward into the air and falling back down. It is like a coiled spring, loaded and then released, propelling the animal forward. The first part of the leap is the fastest as the body accelerates forward, but while it is in the air it begins to slow because of friction with the air. The animal comes back to earth when its velocity is less than the pull of gravity. An animal that is really fast doesn't spend very much time floating around in the air, but keeps the leg levers pulling and pushing against the earth, constantly overcoming gravity and catapulting the body forward.

The greyhound goes *l e a p, l e a p, l e a p, l e a p* along the track, while a horse goes *leap, w a l k, leap, w a l k, leap, w a l k* . . . The rhythm is audible, mimicked by the onomatopoeic *gal–ump, gal–ump, gal–ump.*

If you put a *leap-leap* dog like a greyhound or even a *leap-walk* dog in harness, the instant that all four feet are off the ground, the back strap of the harness will not only stop the flight, but pull the dog backward and off balance and it will fall sideways toward the central gang line. The dog will become unstable. Just imagine having a team of twelve unstable dogs all stumbling and falling at high speed.

Animals that are pulling something must always keep one foot on the ground for stability. Technically, they must be walking, not running. In harness-racing horses, the walk is called a pace or a trot, with two feet on the ground at all times. The runaway (galloping) horse and cart is frightening not just because of the speed but because of the instability, and the fear that the animal will fall.

Sled dogs have walking gaits—pace, trot, and lope. Trots and paces keep two feet firmly on the ground and two feet in the air. These gaits are more powerful than running, less fatiguing, and very stable. In long races like the Iditarod, these are the preferred gaits. But they are slower than the one-foot-on-the-ground lope. The lope allows the dog to stretch farther. It is really a walk that allows the dog to reach out as far as it can physically stretch. The long stride is where the dog generates its speed. Acceleration is still due to the force of gravity, and speed is a function of the length of the stride.

The problem for the sled dog is that, although walking is a controlled falling and recovery motion, the dog is using gravity to pull the sled. If the dog's body weight is not enough to overcome the friction of the sled by just leaning forward (using gravity), the sled will not move, nor can the dog's body fall forward. Thus the total weight of the dog or dogs falling forward has to be enough to overcome the friction of the sled. Consider, for instance, the team of dachshunds harnessed to a sled, all up on their hind feet and poised to fall forward onto their front feet—they are much too small and light to move that sled.

If the total weight isn't enough, then you force the dogs to pull by pushing their bodies forward, using their legs as levers. The heavier the load, the more the dog has to lower its body to increase the leverage. Levering is slow. The more power put into levering, the slower it gets. It is also tough on joints, grinding them against each other. It stresses muscles to the limit. And, it is very, very tough on the feet. When a dog is levering on ice and snow, it has to put enough pressure on the surface to counter the friction of the sled and driver. Slipping results in abrasions; abrasions hurt; hurting dogs don't run. If a dog has to lever, it is working between a rock and a hard place. Races where the team is limited to five dogs are always shorter than the unlimited races, because the fewer dogs have to do more levering with their legs and backs on hills and turns.

By the way, the concept of beating dogs with whips and making them run is from movies and novels. Hurt, terrorized, sulking dogs don't run. Dogs with injuries that hurt don't run. Dogs with sore feet don't run. Dehydrated dogs don't run. A levering dog can't run. The very best drivers are always aware of this and pay meticulous attention to how the dogs feel, mentally and physically. A good driver has to know what the load limits are. A good driver isn't expecting miracles. A good driver knows the physics, at least intuitively. A driver cannot get into the winner's circle without those concepts.

The walking gait, using gravity to provide the main acceleration, is more complicated when you take more than one dog into account. If I have a one-dog team, the one dog has to pull 100 percent of the load. If the dog doesn't have enough weight to move the sled by just falling forward, then I must either get a bigger dog or make the dog I have lever forward. This is when people ask me why I don't get a big, strong dog. The big dog may have enough mass to move the sled by falling forward, so it doesn't need to engage in the slow, painful exertion of levering. Such a dog is called a horse. In fact, British explorer Robert Scott thought he had freight transport in the Antarctic all figured out: he got a bunch of Siberian ponies to pull his sledges. He solved the mass and friction problem, but he created a few very serious new ones.

In order to have mass overcome friction I need either big dogs or a lot of small dogs. Why not use big, strong dogs? Why not a hundred-pound malamute? Why not a two-hundred-pound malamute? How big must this dog be, and how strong? What if I had ten two-hundred-pound malamutes? A ton of dogs leans forward and the two-hundred-pound sled and driver snap forward. What is wrong with this scene?

What is wrong is that after falling forward, each gargantuan dog has to raise its two-hundred-pound bulk back up to full height. Think of doing that thousands of times over a twenty-mile course. When Frank Shorter won the marathon at the Munich Olympic Games in 1972, he sandpapered his racing shoes the night before in order to shave off a few milligrams. Milligrams lifted up and swung forward fifty thousand times during a race amount to a great deal of strength- and speed-sapping force. Big, therefore, has to be strong because big is always heavy. And it takes more energy and more strength to stop heavy from falling forward and then to bring the falling carcass back to its former height—over and over and over.

Can bigger be stronger? Think of a dog as somewhat ball-shaped.

Remember from high school physics that as diameter increases, volume goes up by the cube of that number. And so, for a dog, as the volume is cubed, the cross-sections of muscles and bones are only squared. A man of five feet two inches, fit for the marathon, might weigh a hundred pounds, while one who is six feet two inches could weigh two hundred pounds. That is a 20 percent increase in height and a 100 percent increase in weight. At the same time, the big runner has a proportionately smaller cross-section of bones and muscles. That means he has to work disproportionately harder. It also means that his muscles and bones take a disproportionate amount of abuse. Bigger is not proportionately stronger. People who run marathons, or are champion gymnasts, are small for those reasons, and for several more besides.

Big dogs have more cells to feed, and proportionately less muscle and bone to service the greater volume of cells. The dynamics of really big dogs get out of kilter fast. The reason a Newfoundland likes swimming in cold water is the same reason the brontosaurus liked sitting in lakes. It is not that they have genes for loving water, but rather they are happier when the water supports what their legs have difficulty supporting.

Not only does it support, but the water also cools that mass of cells. Why hasn't someone bred a three-hundred-pound dog? I think such a dog is a physical impossibility, mainly because it would have great difficulty with temperature regulation. Every living cell generates heat, the product of combining oxygen with sugar. Dogs, with their richly vascularized tissue and overhemoglobinated blood with its rich supply of burnable oxygen, have trouble diffusing heat. They are good at burning fuel and generating heat, but they are terrible at getting rid of it. They are not the worst mammal at removing excess heat, but mammals in general are poor at it.

The heat storage graph shows that as weight rises to over twenty kilograms (forty-five pounds), dogs have increasing problems getting rid of heat. And that is just at rest. If the dog moves, the heat load increases rapidly. A fast-running dog produces a phenomenal amount of heat.

Why do dogs have to get rid of excess heat? Dogs, like humans, are homeotherms, meaning they regulate their own body temperature. Temperature is maintained where their cell chemistry operates best. Human brain cells operate best at 98.6 degrees Fahrenheit. If our brain cools to 94 degrees, our brain cells stop the thinking process. Brain

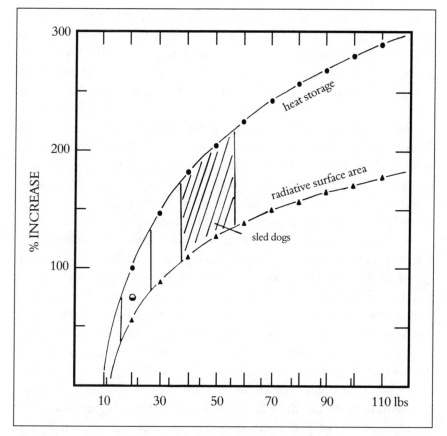

Heat storage capacity. Life or survival is always a compromise. Big and roundish dogs can't help storing heat. Big dogs easily survive an arctic night. But arduous exercise generates internal heat that becomes dangerous if the dog can't get rid of it. Sled dogs are small dogs that are better at radiating heat over marathon distances. (Adapted from Phillips et al., 1981)

chemical reactions can't occur at 94 degrees. If the brain cells get four degrees hotter than 98.6, they begin sending weird pictures, and at 106 degrees the brain cells begin to die.

Normal temperature for dogs is 101.5 degrees Fahrenheit. At 98.6, their cells don't function properly. To try to understand the temperature regulation of running sled dogs, student Dave Schimel and I gave my dogs a snack of thermistors, tiny little temperature gauges that send reports of stomach temperatures to a remote sensing device. We hitched the team to the front of a pickup truck, so we could control and monitor their speed. One day, Dave was driving and I was recording the dogs' temperatures. Suddenly, Midnight's temperature began to drop. "Stop the truck!" I said. "We've got a dog in trouble." We got out and looked the dogs over, especially Midnight. She was a happy little dog and nothing really seemed to have changed. We started again and her temperature continued to drop, and it went way below 98.6. We stopped again. Still nothing seemed wrong with her. She didn't seem distressed in the slightest. Maybe the thermistor is faulty, I thought. We continued on. Dave, at the wheel, was now smiling. "Watch Midnight," he said. I watched as she charged with the others down the snowy trail. Then she leaned over to the side and deftly grabbed a big mouthful of snow. My thermistor was suddenly sitting in a vat full of cold slush!

Humans may be the best mammal of all at getting rid of excess heat, but they are terrible at storing heat. Dogs are just the opposite—terrible at getting rid of heat but experts at storing it. The balance point between trying to retain heat and trying to lose heat for humans is an ambient temperature of 70 degrees Fahrenheit (depending on size and shape, of course), and for smooth-coated sled dogs it is 60 degrees. As a dog driver, I wouldn't train if the ambient temperature was over 60 degrees because the dogs would be at risk.

A look at the graph again shows that the little dogs had less of a problem with heat storage than the big ones. It is probably significant that racing sled dogs weigh fifty or so pounds, just at the dividing line where heat storage capacity starts to increase enormously with increasing size.

Big dogs have proportionally greater cell volume and proportionately less surface area to radiate the heat. Since dogs don't have sweaty, bare skins to radiate heat, evaporative cooling is not an option for them. Panting hard cools the lungs and brain, but the only place a dog sweats is through the pads of its feet. The pads just don't have enough surface area to make them effective radiators. Dogs remove excess heat from surface areas just like a radiator. They dilate the veins and arteries under the skin. Hot blood is exposed to this cooler surface, which radiates the heat away from the body. Of course, the surface is covered with fur, which makes it harder to radiate the heat. And the bigger the dog is, the worse the heat-load problem. Remember the volume is going up by the cube while the surface area squares. Big dogs have proportionately less surface area to radiate heat from.

I can't say this often enough: dogs have trouble radiating heat. Every time I see a fat dog I cringe, thinking how it must suffer even with modest exercise.

The written standard for Siberian huskies states a maximum allowable weight of sixty pounds. Why is that? Why didn't dog drivers develop bigger, and hence stronger, Siberians? Why would it be cruel to run a malamute at racing speeds? Why doesn't it make sense to breed a hundred-pound wolf to a sled dog? The answer to all, of course, is because the heat load would be too much for them.

Doing research on temperature regulation in sled dogs, I was amazed to find rectal temperatures of 108 degrees. If they didn't have such an amazing mechanism for dealing with this heat, they would not be able to run so fast. But dogs can cool their brains with their tongue and nose. In other words, they can maintain a normal brain temperature while their body temperature climbs. The highest rectal temperature I ever recorded was 109 degrees, and Dolly was still on her feet and running. But that is dangerously close to tissue death. If body cells reach toward 112 degrees, they begin to die.

Big, strong malamutes become overloaded with heat very quickly at racing speed, and can literally start popping blood vessels in their brains. The cross-sectional dimension of their tongue is not much greater than that of my "puny little mongrels," and it has to cool a much larger volume of cells. That is why I'd rather take a chance with dachshunds. I probably couldn't get them going fast enough to hurt them because of heat buildup.

It turns out that the *leap-leap, leap-leap* greyhounds have the same problem—too much heat-storing capacity at sixty-five to eighty-five pounds. Their streamlined appearance means their surface area is greater than that of a more normally shaped dog. Maybe their radiatorlike shape was originally an adaptation to desert life. Maybe running fast is just one consequence of being adapted to very hot conditions. Maybe a twenty-five-pound tropical village dog is also an adaptation to tropical heat and humidity.

Why is it that sixty-pound greyhounds are so fast in short races, and don't have overheating problems? Greyhounds are sprinters and sled dogs are marathoners. Sprinters (dogs or humans) run their entire race on liver glycogen, sugar mobilized from the liver. The trick in sprinting is to have long legs and a big liver. Move the legs very fast, with energy from burning sugar released by the liver. Big livers store more sugar than small livers. Once the liver glycogen is used up, which can happen in less than a minute, the dog or person has to switch to fatty acids for energy, and switching from one to the other takes a little time.

In the ten seconds it takes a human to run one hundred meters, or the thirty seconds it takes a greyhound to run five-sixteenths of a mile, the sprinter hardly needs to take a breath. At that speed a person or a dog, running out of liver sugar before the end of the race, would just collapse on the course.

There is a nice size parallel between human and canine sprinters, and human and canine marathoners. Human sprinters (male) weigh about one hundred and sixty-five pounds, which is 30 percent bigger than a good marathoner of one hundred and twenty pounds. The same ratio exists between a greyhound sprinter and the distance-running sled dog. That is why the greyhound has no stamina. It runs out of energy quickly.

What about heat buildup in the bigger sprinters? It's usually not a problem, because the race is over before it becomes dangerous. In races that last just a few seconds, the sprinter can simply let the heat build up. In contrast, marathons take over two hours, and heat load becomes just as much trouble for the human as the dog. The difference is that the mostly naked, perspiring human is significantly better at getting rid of that heat and can run those races in higher ambient temperatures. In the Introduction, maybe I should have said that sled dogs are the fastest animal in the world at distances over twenty-six miles at temperatures under sixty degrees. My guess is at ninety degrees, humans might edge out dogs. But I wouldn't run a dog at that temperature to find out.

If big dogs can't run marathons, then why not run Chihuahuas? They shouldn't have any problems with heat load (and they don't). Chihuahuas also should have proportionately the biggest-diameter bones and muscles (and they do). But no one asks a sled dog driver why he doesn't run Chihuahuas. The answer is obvious. Chihuahuas may be proportionately stronger for their size, but "for their size" is the crucial statement. Think about them leaning into a harness, bringing their weight to bear to pull the sled. Think about them, given the length of their legs, levering a dog sled forward.

If it wasn't for the heat and weight loads, big dogs (*walk-walk-walk*) would really be a lot faster than little ones. The bigger the dog, the faster it can run. Why? Because with every single stride, the dog covers more ground. All other things being equal, the animal with the longest back, the longest legs, or simply the longest reach is going to win. Stride for stride, the longer dog will pull ahead as it covers more ground with each step. I preferred smaller dogs (forty-pounders) for the warm New England races. But I lost a lot of races because those little dogs didn't have the reach.

Reach is, of course, more than just overall bigness. The positioning and the shape of the shoulders make a difference in how far the dog can extend its legs forward. Some dogs don't have enough space between the top borders of their shoulder blades. When their front legs stretch forward, the scapulas move together at the top. When they touch, the front legs cannot go any farther forward. If you force a dog to run too fast, it overreaches, the scapulae keep rubbing and bumping together, creating a sore spot, and the dog "cripples up." Dogs that hurt don't run well, if at all.

One of my best leaders was too tight in the shoulders, which he compensated for by pumping with his wrists. This is not as fast and takes more energy simply because there are more moving parts. Moving parts wear out. With more moving parts, Perro could not run long, fast races. With such a dog, one has to be careful not to hurt him by pushing too hard and forcing him to overreach or move his wrists too many times. (He was a super dog. He could adjust his parts to achieve his maximum speed. He must have done a lot of running on what people call "heart.")

The same considerations are true for the pelvic girdle. If the pelvic bones are at a steep angle to the backbone, the rear legs can tuck up underneath the dog. The dog can reach a long way forward with its hind legs. If the pelvic girdle runs straight off the backbone, then the legs are hinged straight out the back, like a seal's. If I see dogs running with their tails up, I know they can't be reaching forward with their hind legs.

Dogs and coyotes sweat through their footpads. None of the wolves I tested could. The hot dog sweats onto the little hairs between its toes, and if it is running on snow, the sweat forms ice crystals. Ice between the toes cuts the feet, and sore-footed dogs won't run. Good drivers watch the snow conditions carefully, and prevent the team's feet from becoming icy and painful by outfitting the dogs with special booties. Of course, this does affect how fast the dogs can run, but preventing sore feet is more important.

Perro. The best leader ever. Perro wasn't a great-looking border collie (he even had papers), but he was a great sled dog and fathered other sled dogs. I have worked with a lot of great dogs, but I think this was my "once in a lifetime" dog. I got him from my statistics professor, Jim Denton, because he couldn't keep Perro from chasing cars all day long.

All these anatomical differences between individual dogs—shoulders, pelvic girdles, leg length, back length—result in different gaits. The differences are most obvious between breeds, but no less significant between individuals.

The whole point is this: the shape of the animal provides the limits of its behavior (its motion through space and time). The shape of the dachshund limits the speed because it has no reach. A malamute can't run far because of heat buildup. In spite of the traits of other breeds that might indicate more speed or more endurance than that possessed by a sled dog, no breed possesses more of the consonance of characters that results in a good, fast racing sled dog.

It should be obvious by now that it is not the standing shape that is important. Standing in a show ring reveals little about a Siberian's running shape. Charlie Belford had a wheel dog—the sled dog that is hitched just in front of the sled—by the name of Sammy. Sammy appeared badly built when standing in the yard, but he was beautiful at

Matched gaits. The team's running conformation is really good. Note how the leaders and many of the other dogs are all in step. Note their waving tongues. In order to run long distances they must keep their body heat, especially their brain temperature, down.

Wehle pointers. This is Bob Wehle and his nationally famous team of pointers having fun in the off-season. When a breed of dog hasn't been selected for a task they look a little comical attempting it. There is a lot of waste motion here and in a really long event these dogs couldn't compete with a modern racing sled dog. (Courtesy of Lorna Demidoff)

full speed. His cow hocks and roached back disappeared, transformed into the smoothest, most energy-efficient gait you would ever want to see. The faster he went, the prettier he got. He always reminded me of the baseball player Mickey Rivers. I thought Rivers moved awkwardly when he walked, but he ran beautifully, and very fast.

What an important point this is. I want to see the shape of the dog at twenty miles an hour. I don't care about the standing shape, or what the shape is at twenty miles an hour when the dog is free-running, chasing a ball. I want to see the shape of the dog putting pressure on the back strap of its harness, at speed. That is the only meaningful shape for a running sled dog. That is why it is hard to judge conformation at a dog show and why serious sled doggers have little interest in a dog show. It is not how they want to see a dog.

Another way to say this is: standing is a behavior. Running is a behavior. If I want standing dogs, I select for dogs that stand easily and economically. I select for dogs that like to stand and have a beautiful standing shape. As a sled dog racer, I am not much interested in a standing shape.

THE SHAPE OF THE TEAM

Each dog on a team has a conformation. But, to race successfully, the team as a whole must have a perfect shape.

Extending forward from the front of the sled is the gang line. All the dogs are hitched in pairs along the gang line. When everything is perfect, the gang line is exactly straight. There are no bends in it. If the gang line is crooked, then energy is being lost. In order to achieve a straight gang line, pointed directly at the finish line, all the dogs should be exactly the same height, the same size, and work synchronously together. Ideally, there wouldn't be any bends in the trail—but that would take a lot of the romance out of racing.

At winning speeds, the sled dog's body has to put some energy into the back strap. To pull me (two hundred pounds) on a sled (thirty-five pounds) at twenty miles an hour on the flat, with no wind, and snow not so cold that it drags, takes eighteen pounds of force. Divide that by sixteen dogs, and theoretically each dog should put a continuous 1.125 pounds of pressure on the back strap. It doesn't sound like much.

But 1.125 is the theoretical figure and assumes a perfect system. And—no big surprise—it is not a perfect system; every tiny imperfection costs each dog more energy. Whatever force the dog has to put into the back strap has to be subtracted from the energy it puts into speed. If a dog is running absolutely as fast as it can go, it should not have any strength left over to pull with. As the dog increases speed, it puts less and less into the back strap, until at the top speed nothing is left over. As the speed increases from zero, the amount the dog can put into the back strap decreases rapidly.

Note that the force:speed ratio is curvilinear. In an ideal sled dog world, all the dogs would have the exact same curve. But in the real world, the top speed and shape of each dog's curve will be slightly different. If the shape of every dog's curve is different, then each dog will be putting a different amount of energy into the back strap when it is running at the same speed. That means the gang line will be crooked and energy is lost. The conformation of the team dictates how crooked that gang line is.

Ideally, a driver could run in pairs those dogs whose running conformation matched each other at racing speed. That is the point (**x**) in the force-speed graph where the two curves cross. A fast-running sled dog team, of course, needs to have dogs of fast-running conformation, but a team isn't made up of one dog. Nor is a team made up of a random collection of well-shaped dogs. In order to be a winning team, it has to have its own conformation that blends together. And this specificity differs between drivers. Attention to these details is what often separates the really top drivers from the almost-top ones.

The way this works is to match dogs with similar gaits and speeds in pairs along the gang line. When putting the team together, drivers try to match dogs that have nearly identical force curves. That means I have to know what speed I am going to average over the course, which means I need to know how hilly it is, how many curves it has, what the snow temperature is, and all the other factors that determine how fast the trail is.

When I watch sled dogs running, I see each segment of their stride separately, and it is these segments that I have to match between dogs. As I mentioned earlier, running for sled dogs is fast walking, gravity pulling the animal forward and down. Here is where a good imagination will help. Imagine a dog standing on its hind legs, humanlike, with its back strap attached to the sled. Now the dog leans forward and falls,

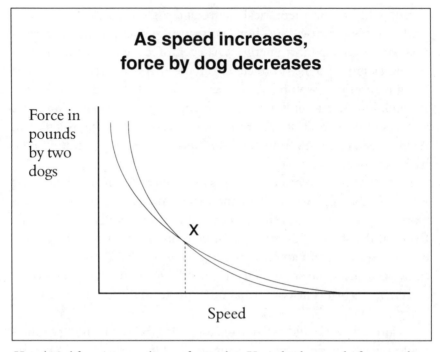

As speed increases, force by dog decreases

Force in
pounds
by two
dogs

X

Speed

Hypothetical force times speed curves for two dogs. X marks the speed of a race where these two dogs would gait well together.

keeping its hind feet in place. As it falls forward, the sled has to move forward an equal distance to the arc of the fall. This dog is putting no more energy into the system than its own falling weight. That's perfect. A simple lever device, no moving parts—efficient and stable. All the dog has to do is gather itself up to full height and fall again.

Here is a dog standing vertically on its hind feet, attached to a back strap, attached to a sled. But where on the dog is the back strap attached? To its head? Its shoulders? Its waist? Just above its hocks? The lower the attachment, the more force the dog puts into the sled, but the less distance the sled moves. If the dog was perfectly rigid and the back strap was attached to its hock, just by falling forward the dog would put a tremendous force on the sled. That's the good news. The bad news is the sled would move only a couple of inches. The other bad news is that unless the dog has good, heavy bones, it will break its leg, because the same force put on the sled is also put on the hock.

How a team runs: good team, bad team. Because of their different sizes, gaits, and poor training, the bad team can't run efficiently. (Adapted from Phillips et al., 1981)

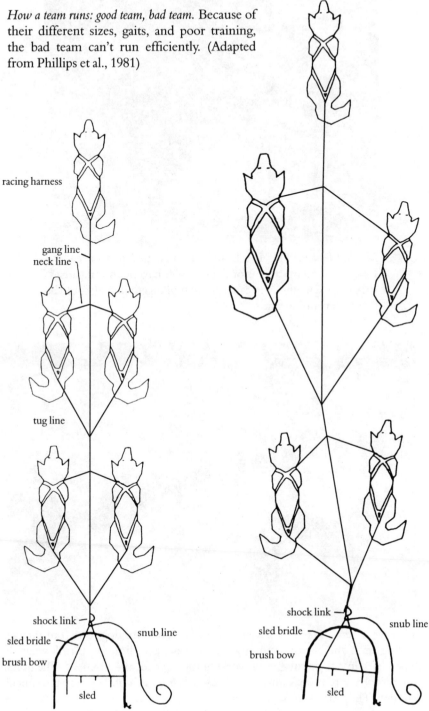

racing harness

gang line
neck line

tug line

shock link
sled bridle
brush bow
snub line
sled

shock link
sled bridle
brush bow
snub line
sled

Matched leaders. This is a beautifully matched pair of leaders. Dogs with identical conformations will have identical gaits; if they like each other and can learn to pull together, they could be a perfect running pair.

Team conformation (and a tired dog). Here is a wonderful dog team, but they are getting tired and are overrunning themselves just a little on this downgrade. They have lost their rhythm of working together for the moment. The dog in the inset has tired psoas muscles and can't bring his hind legs forward quickly enough. They just float back and up.

If I raise the back strap, the falling dog is not able to put as much force into moving the sled, but the sled moves farther. The analogy isn't so silly as it might sound. The driver is always trying to figure out just such equations. A big factor of the running equation is how and where that harness fits. Freight harnesses pull from lower on the dog than racing harnesses. A good racing driver pays meticulous attention to how the harness fits. It should pull from as high up on the dog as possible, and should fit in such a way that it doesn't restrict any moving parts. The trick is to get all the force centered on the dog's breast-bone—which in a perfect world doesn't move, doesn't articulate at all. A horse collar doesn't work on a dog because it would ride on the dog's forearms, which are what the dog is reaching forward with. If the straps from the breastplate are loose and run over the dog's shoulders, which are moving far every time the dog reaches forward, then the dog is pulling with its moving shoulders, the harness chafes the shoulder, and the dog wears out. A worn-out dog hurts and won't run.

One more time. Here is the dog standing on its hind feet and falling forward, with the back strap pulling from a point that pulls right on the breastbone. But now the sled is too heavy and the dog can't fall for-ward. What to do? Get a bigger dog? No—heat load—won't work. Okay, so I try hooking together two dogs standing up and falling for-ward. This exerts more force. But each dog has to fall forward at the same time. If one falls forward and nothing happens and then the other falls forward, still nothing happens.

So I now have two dogs "trained" to fall forward at exactly the same moment. The sled still doesn't move. I try four dogs, then six, then eight, and at some point the sled begins to move. Not only that, but if I have enough dogs falling forward at exactly the same instant, the sled will move forward at twenty miles an hour. For me, at my weight, twelve dogs is the magic number.

In order for twelve dogs to be my magic number, though, every dog has to be exactly like every other dog on the team. Here is why.

RUNNING IS SOCIAL BEHAVIOR

If I had a one-dog racing team (standing up and falling forward), that one dog has to pull 100 percent of the load. If I had two dogs standing on their back legs falling forward, I might assume that each would pull 50 percent of the load. But that is physically impossible. Each dog would have to be identical in size (improbable), fall in an identical arc (improbable) at the identical time (improbable), and occupy the identical place (totally impossible).

My original model assumes that each of the dogs is standing on its hind legs hitched to the sled, and falling forward at the exact, same instant and falling through the exact, same arc. If one dog has a bigger arc, it is not putting as much pressure on the back strap—even though it pulls the sled farther.

Trying to keep in mind all these factors and balance them with the available dogs, is the challenge of putting together a top racing dog team. Since no two dogs can be exactly the same size, then not only will they fall through different arcs, but the falling through different arcs will take different times. So the distance traveled with each fall is different, and the time of the fall is different, and therefore the gang line has to bend.

The pressure of falling not only pulls the gang line forward, but pulls it toward the dog. Since no two dogs can occupy the same space, no two dogs can pull in the same direction. Each dog is pulling the sled in a different direction. If the back strap is at forty-five degrees to the gang line, then 50 percent of the forward motion created by the dog is lost to the system because of that angle. It is simple physics.

In theory, each dog of a two-dog team would pull 50 percent of the load; but in actuality, on a six-dog team each dog is pulling 50 percent of the load. If I were racing a six-dog team, each of my dogs would be putting a continuous nine pounds of force into the system—18 percent of their body weight.

Twelve dogs works best for me. Once I get over twelve, each new dog equals all the inefficiencies it brings to the system, adding nothing to the improvement of pulling the sled and me. With more than sixteen dogs, the system becomes so unstable that a driver can quickly lose control. The reason many of the top drivers run sixteen dogs is so they can, if necessary, drop a pair of dogs on the first and second days of a three-day race, and race the third day with twelve dogs. Once a race starts you can drop dogs but you can't add fresh ones.

Because the team is a social system, requiring each dog to participate in unison with every other dog, drivers strive to put together a matched team. Every dog should be built exactly the same as every other dog. The dogs running on the left side of the gang line should be mirror images of the dogs running on the right side. If the dogs on the left side are left-footed (they lead with their left foot), then the dogs on the right should be right-footed. All the dogs should be gaited exactly alike; all dogs should have exactly the same reach; and all dogs have to be trained to run together—work together—to synchronize their motion. Each dog can learn to run with its partner, such that they can adjust for speed/power curve differences by changing the angle of their back strap. With all its imperfections, a good team can keep that gang line straight. They must be able to sense when the gang line is straight, because they are going just as fast but using less energy. (The principle holds for walking hounds, too. Uniformity of size and shape is necessary for a good hunt.)

The driver (and the master of hounds) is like the coxswain on a sculling crew, keeping everybody *exactly* synchronized. If they get out of synchronization, dogs can't work as well and will lose the race. Drivers have many little tricks. Lengthening the harness back straps decreases the forty-five-degree angle to the gang line. But lengthening the back strap too much increases the risk of serious tangles. On curves, when the front dogs have gone around a corner and literally can't put any forward force on the gang line, the rear dogs take all the pressure. A good driver always runs the corners. This takes some of the pressure off the wheel dogs (right in front of the sled), which have more of the load but have to keep the sled on the trail as they get pulled into the corner by the dogs ahead. On a narrow, winding trail, long teams become big problems, and the team needs to be shortened. Fewer dogs and a shorter gang line and back straps do the trick. Drivers adjust the team and the lines, and therefore the running conformation, to each trail.

The story is not quite over. It is idealistic to think every single dog could be trained to run at exactly twenty-three degrees from the gang line. One dog may run close to the gang line, at twenty-three degrees, but its running mate cannot stand to be that close to another dog and so runs forty-five degrees from the gang line. Now, if both dogs are reaching and pulling the same, one dog is losing a quarter of the forward thrust and the other is losing half the thrust. That means the gang line has a bend in it. If the gang line has a bend in it, the sled is being pulled off the center line and wants to rotate, thus introducing another inefficiency—torque.

One of the big lessons in training dogs is to train the pairs of dogs to like each other the same amount so they pull on that gang line an equal amount. That is why a new dog is so hard to introduce to the team. Until that new dog gets to know its partner and the other team members, it can be standoffish, and that puts a bend in the gang line. Sled dog drivers know about critical periods of social development. They raise puppies in play groups, perhaps feeding them separately so no quarrels develop over resources. They don't let them fight as puppies, which means less likelihood of fights as adults. Playing among pups leads to dogs that want to be with other dogs.

It is, of course, impossible to get twelve or sixteen identical dogs, all of which like one another the same amount. Thus, physically matched pairs are usually the closest one can come to perfection. Much of training is figuring out the pairs. I try to find two dogs that are gaited alike and then teach them to run together. I once had this super dog named Pang. He was the fastest dog I ever had. But every week it was the same dilemma: who to gait him with. Finally Charlie Belford took him, but he had the same problem. The dog was simply too fast. Even Pang was unhappy, for he had to work harder than his partner simply because he had more reach. Pang was always ahead, and his partner and the sled would be jerked toward Pang with each stride. The sled followed Pang to the right and was straightened out by the dog on the left, and then restraightened by Pang on the right and so on down the trail. Thus the sled went inefficiently down the road, zigzagging back and forth.

If I ran Pang on single lead, then the whole team had to run at Pang's speed—overreaching their stride—or Pang had to slow his speed. He must have felt like an adult walking with a little child. Not only does the child have to overextend his gait to keep up, but the adult has to shorten his own stride. Both individuals are uncomfortable. If Pang slowed down to the slower pace, he looked like Mickey Rivers trying to walk. He was too much dog for me. If only I could have had a whole team of Pangs, I could have been the unbeatable champion. But selling him was the best strategy for my team.

A dog driver is a choreographer of all these behaviors. As the race is run, dogs tire at different rates and lose the rhythm. A tiring dog has to exert more force to stay at racing speed. The rest of the dogs have to adjust to the adjustment. A driver will have to work hard to get tired dogs "back in there." He talks encouragingly to them, or stops for an instant to let everybody start on the same foot again, or even rearranges the team, matching two tired dogs together to keep it balanced. If the tired animals aren't costing very much, that is, they are just not pulling, then it may be best not to get on their case too hard because vocal force might disrupt the rest of the organization by encouraging the still-effective dogs to work harder, thus throwing them out of rhythm with their companions, and then you're in a mess. If a tired animal is dragging back on its neck line and being towed, it may be time to load it on the sled.

THE SOCIETY OF A SLED DOG TEAM

Running together as a team is a social event, a system of togetherness. How well that social system works depends somewhat on the collection of individual talents, but even more on how each individual's running shape and size mesh with that of other members of the society. One dog can't pull a person at racing speeds, nor can two, three, or perhaps even six or seven. One dog could never be big enough to pull a sled far and fast because of physiological limits on the size at which the animal can work. The workload must be divided among a number of individuals in order to accomplish the task. And still, how well the individual dog works depends not only on its willingness and ability to contribute, but on the availability of a good place for that individual's conformation in the system. Each dog might have the ability to run at speed, but that isn't enough; each dog must run in synchrony with another dog, in a team of dogs. Talent means having a running conformation that can do that. Perfection is not having superior talents, but having identical talents, whatever they are, among the dogs.

This description of a social system is very different from the way dog teams are often described. The pervasive cliché is that they are like a pack of wolves, with a leader which dominates the pack. The analogy then designates the driver as the supreme pack leader, the so-called alpha dog, exercising his will over the pack and forcing them to run with threats of physical violence. The driver, it is believed, forces these subordinate animals to be submissive, wielding a whip, which he cracks in the air to make the dogs speed up. The mythology becomes complete with references to northern people staking out their female dogs so local wolves can breed to them. The rationale is that breeding to wolves gives the sled dogs stamina.

Nothing could be further from the truth.

Maybe some people did breed their dogs to wolves and then had to beat them with whips just to keep them from killing one another when they harnessed them. It is not intuitively obvious that breeding to hundred-pound wolves would introduce a host of counterproductive elements into the team. But stamina would be reduced. Dominance hierarchy behavior would occur, and disrupt a coordinated effort. Synchrony of running would vanish, because wolves are independently minded animals that think more about their personal space than dogs do. Wolves typically react to commands by assuming submissive postures, or quit on a whim and then sulk. I can't think of a single trait possessed by wolves that I'd want on a dog team. Ken MacRury, who has studied northern dogs for years and has raced with Inuit dogs, says the Eskimos he knows just laugh when someone suggests breeding dogs with wolves. They don't believe those stories either. Some Alaska drivers attempted to add wolves to their teams, largely without success. Jack London's fictitious lead dog, Buck, dreamed about being a real wolf, and in the end left the world of man and reverted to the wild to lead a wolf pack.

To me, this kind of imagery is not just fiction, but awful fiction. It seems to me there should be a touch of reality to a romance. London's story does no favors for dogs, or for wolves. The idea that I would go out and pay five thousand dollars for a lead dog, bring it home, and let it fight the rest of the dogs to see if they accept it as the alpha male and leader has to be hilarious. The last thing any dog driver wants is a dog fight. Not only can valuable animals be hurt, but animosities between individuals would constantly stress the team effort. I don't want any dog to feel bad about its rank on the team, or continually test its position. I don't want the dogs submissive to me. Imagine them all on their backs, peeing in the air every time I showed up.

On a twelve- or sixteen-dog team, the leaders are usually paired. The leaders can be paired as males or females. What does that do to the theory of the alpha dog? Two female alpha dogs? A dog team with good depth has many leaders. It has alternates that the driver can substitute up front to replace animals too tired to keep a winning pace.

Dogs are not wolves. Dogs are not running as a pack. A pack is about chasing something. Sled dogs are running because other dogs are running. They are motivated by something the animal behaviorists call social facilitation. There is a rhythm to their run and they can hear that rhythm and they run to it. When you stand on the back of a sled, you can feel it. It is powerful.

Each individual dog is a character. Some are happy, while others can be grumpy. I had an older dog, Red Raino, that I got from Charlie Belford. She hated training but loved a race. One time I got stuck at the annual race in Rangeley, Maine, for a leader, and I put her up front. I would have won that race, but five miles from the finish line I couldn't get her to "haw" and head for town, and the team took a different trail. When I told Belford, he said, "Oh, that's the way we use to go in the old days."

I had standing orders from my family not to leave any dog at home on race weekends. The left-behind dogs would howl, bark, and cry all the time I was gone. The first time I ever did it, I left Rena, who had sore feet, to give her the weekend off to rest and heal. I never did that again. When I got back Sunday night, she was hoarse. Nothing but squeaks came out as she continued to bark—but she was very glad to see me.

The structure and behavior of a dog may superficially resemble that of wolves, but in fact, to focus on similarities does the dog a great disservice. Sled dogs are an evolutionary advancement over wolves. Sled dogs are as close to an evolutionary perfection as you can get. They do something better than any other organism. Why would anyone want to mess them up by breeding them to wolves?

THE VALUE OF THE BREED STANDARD

Looking at a well-bred dog or a carefully managed team is like looking at a great boat. The lines may not be intrinsically beautiful, but rather the beholder recognizes and appreciates just what that shape can do. The more experience the beholder has, the better the recognition of the great performance possible with that shape.

Running is a behavior. A running team is exhibiting complicated behavior. These dog teams have a special shape that allows them to behave better than all other creatures at that task. Sled dogs are not an ancient breed, nor a direct descendant of any breed. The running of dogs is not an ancient art that we are copying and trying to preserve.

The dogs and drivers of today are the most complex, the best that ever existed.

I remember a dark gray dog named Tony whom I bought from Charlie Belford. At the end of a race the dogs would be tired. Some would lie down, stick their noses in the cool snow, or lick a paw. Each would attend to its personal needs. After a short while, Tony would push his way down through the snow and then come up standing, shake from one end to the other, and with this totally arrogant "conformation" he'd say to me, *Well, is that it?*

Tony and I may not have enjoyed a true symbiotic relationship, but we had a common language, one that is well understood by people who are close to dogs.

CHAPTER 6

Behavioral Conformation

HERDING DOGS, RETRIEVERS, AND POINTERS

Mutualism is the type of symbiosis that benefits both species. But it often implies that there is an evolution of form, where both species adapt to living together. Hummingbirds evolved long bills that are good for one thing—extracting nectar from a trumpet-shaped flower while simultaneously pollinating the flower. Both bird and flower have specific adaptations that the other depends on for its survival.

I'll now look at dogs that perform specialized tasks with humans. As in sled dog racing, these tasks could not be carried out without the dog and human working together. Neither species could accomplish the objective alone. This is the classic symbol of a person and a dog as inseparable companions.

It is not, however, a true biological mutualism, because the survival of either species is not dependent on accomplishing these tasks. At best, it is a convenience for the humans. Most shepherds in the world don't use herding dogs, and indeed, the work that goes into training and using a border collie is more trouble than it is worth for most farmers. Certainly there are applications in Australia and New Zealand where a trained kelpie or huntaway is worth its weight in gold, but human survival is not dependent on the animal's performance. In the great transhumant migrations, shepherds can move millions of sheep a year without the aid of a herding dog.

The use of sheepdogs, pointers, and retrievers is mostly for fun. They are used in sport trials, often by professionals who pit themselves against one another before an audience for the purpose of testing their skills as dog breeders, trainers, or trialers. Who can develop and display the best dog? Trialers win prize money and the big winners sell more puppies. But mostly their incentive is the love of watching their dogs in competition.

Unlike their masters, however, dogs change form as a result of the tasks they perform. Starting in the long–ago past, village dogs raised as walking hounds were selectively bred for a particular kind of behavioral display. This display has a form. It is a stylized display, an exaggeration of behavior. Pigeon fanciers develop breeds like rollers and tumblers that have exaggerated flight patterns. Horse people develop breeds and training techniques, such as dressage, which are lovely to watch. And the dog people outdo them all by developing hundreds of breeds, each with some dramatic display of some bizarre conformation: red setters, golden retrievers, English pointers, pit bull terriers, Hungarian pulis, and, my favorite, the border collie. (I know that many people would disagree, but I think the border collie has the most delightfully bizarre behavioral conformation of all the breeds.)

The consummate skill of the sled dog, described in Chapter 5, is due largely to its unique physical conformation at running speed. "Running is complicated behavior . . ." I said. The lesson was that the winning behavior was a function not only of the individual team dogs' shapes, but the team's continuing *physical* shape over a racecourse. In this chapter, we look at breeds of dogs where the individual's *behavioral* conformation is directed over some course.

Behavioral and physical conformation, in the final analysis, are one and the same. As a behaviorist looking at genetics, I understand that there is no difference between a behavioral conformation and a physical conformation. The sled dogs excelled at one behavior, running, which was the result of many elements of conformation. Now we will look at dogs whose physical conformation is not an obvious asset to their *specialized behavioral conformation.* The accomplishment of breed-specific tasks is a sequence of exact and stylized movements. It is like a ballet, where the performer must display a number of designated moves in some continual way.

Behavioral conformation here is a description of the behavioral shape—how the dog moves—when a dog is working. In this chapter, I want to show that just as sled dogs inherit their physical shapes, so do herding dogs, retrievers, and pointers inherit their behavioral shapes. Just as the sled dog driver choreographs the running shapes of his racing team dogs in order to attain the perfect performance, so does a border collie handler arrange the behavioral conformation, the innate behavioral shapes, of his dog, to exact the placing of the sheep in a pen in the shortest time.

First we must understand what behavioral conformation means, by looking at behavior genetics.

If I were judging a German shepherd at a dog show and it had cow hocks, I would give it low marks. As a dog show judge, I would say the dog had poor conformation. As an ethologist, I would say this dog was standing improperly. This dog has a poor behavioral conformation.

Standing is a behavior. This animal is standing incorrectly. The cow-hocked dog has assumed the wrong posture. Believe it or not, this is not nitpicking or a semantic obfuscation. It is one of those places in the biological world where people are constantly confused, and constantly misuse terms. Example: I go to the doctor and complain about my arthritis. The doctor says to me, "Do you realize you have poor posture?" And I say to her, "I don't have poor posture, I have poor conformation." I have good posture given my conformation. She assumes that if I could behave in a standing-straight manner, my health would be normal. I find it uncomfortable to stand straight, and most of the time I can't get into that position. When we say that a dog has poor conformation, we're giving him the benefit of the words, because we are assuming he cannot stand straight.

If I am a judge at a border collie trial, and a contestant nips at the sheep's heels, I take off a wee point because the dog has taken on the wrong shape. The dog is displaying the wrong behavioral conformation for a sheep-herding dog. As judge, I assume the dog cannot assume the correct posture and perhaps should be disqualified as a border collie. If I were a judge of Queensland blue heelers and I observed a dog that didn't bite the cow's hock, I would deduct points for what is inappropriate behavioral conformation for a heeling dog. If I judged that the animal could not assume the nipping posture, I would assume that this was a genetic—meaning structural—defect, and disqualify the dog.

This is a very different approach from the more common albeit anthropomorphic view of dog behavior, where the observer assumes the dog learned to perform the behavior. What I hope to show here is that each dog within a breed has an innate behavioral shape for each portion of its performance. What a judge is doing is giving a dog a score for the shape of each posture. The border collie gets scored on the shape of the outrun, then the shape of the clap, and the shape of the chase, and so on. The judge has in his mind a gestalt for what that shape should be, just as a show judge has an idealized image of what a perfect dog within a breed should look like.

In one case the judge is looking at the physical form, in the other the judge is looking at how the form moves through time and space. Both judges are actually doing the same thing; they just describe it with a different vocabulary.

In his book *The Intelligence of Dogs,* Stanley Coren ranks cavalier King Charles spaniels forty-fourth, Siberian huskies forty-fifth, and dachshunds forty-ninth on his list for "obedience and working intelligence." He ranks border collies first, golden retrievers fourth, Chesapeake Bay retrievers twenty-seventh, and English pointers forty-third.

Could I teach dachshunds (#49) to pull dog sleds? They are at just about the same level of intelligence as Siberian huskies (#45). One of my best sled dogs was a purebred border collie, first on the intelligence scale. I bet I could develop a whole team of Afghan hounds, last on the list of 133 breeds. I bet I could teach Afghan hounds to pull a dog sled better than dachshunds. But how can that be since the dachshunds are so much more "intelligent"?

Whoa, you say, pulling a dog sled has nothing to do with intelligence. Each of those breeds has some unique and different conformation. Some of those breed shapes are much more adaptable to learning to be a sled dog than others.

Exactly!

Remember the story in Chapter 4 about Brian Plummer in Scotland, who built his dog-training career on the fact that he can teach any breed any task? Brian believes that nurture is more important than nature in determining intelligence. I spent a delightful day hunting rabbits with Brian and his King Charles spaniels (#44).

"Why did you pick this breed?" said I.

"Because these dogs have been the epitome of the housebound, nonworking pet dog for centuries," said he. "If I can train them to hunt rabbits, I make my point: I can train any breed to do anything." And sure enough, after scampering over hill and dale (actually, the neighbor's lawn), holing the rabbit, and driving it out with a ferret, the King Charles spaniels killed it.

Then we went down to the river and put his golden retriever (#4) through the fetch, retrieve, and deliver-to-hand routines. Nice dog! She would work all day long and never quit. I asked Brian if he could train his golden to herd sheep, and he responded instantly, "No problem. It would be easy to train a golden to herd sheep."

"Could you then take that dog into a sheepdog trial and win?"

"Oh no!" he said without hesitation. I knew I was talking with a person who understands behavioral conformation.

"Oh no!" Just like that. The reason he cannot teach the golden to win a sheep trial is the same reason I cannot teach a dachshund to win a sled dog race. They are the wrong shape—the wrong conformation. The dachshund has the wrong physical conformation and the golden has the wrong behavioral conformation to herd sheep.

Like Plummer, Darwin and Hal Black and Jeffrey Green (see Chapter 4) tended to think any dog could be nurtured as a livestock-guarding dog. Raise the pup during its critical period according to a recipe (nurture). If you follow the recipe exactly you should get the same results every time—but only if you use the correct ingredients. With the livestock-guarding dogs, we want them to grow up and be social with sheep; thus we shape their developing brains by "growing" them in the correct environment. The same is true of sled dogs. We want them to be social with other sled dogs, so we raise them in big pens with other sled dog puppies and let them develop playful social skills with other dogs.

But now I have added something else to the recipe—the right ingredients. Each dog is genetically limited in how much the environment can modify its shape. If I raise my Pyrenean mountain dog in a penful of sled dog puppies, it might be social with other dogs, but it won't look like a sled dog, nor will it have the talents of a sled dog. If I raise a border collie on the Chesapeake Bay, it will not develop the talents of a good retriever, even if it is the smartest dog in the world.

That does not sound like an earthshaking discovery. I have trained border collies to retrieve ducks, but it is just fun, perhaps an interesting novelty. The border collie isn't as big and its mouth is the wrong shape and it does not have the body volume of a swimming dog. It gets cold. It lacks the appropriate physical conformation. But what if I took a German shepherd or some other breed that was about the same size and still couldn't get it to perform like a good retriever? I'd have to conclude it's not only the size that's important. Rather, there is something about the shape that underlies the perfect performance.

Shape of the perfect performance? It is the shape of the brain in the breed that is the underlying cause of the unique intelligence. Intelligence is not more or less in each breed, but rather, each breed has a different kind of intelligence.

Could I teach a border collie to point birds? Yes. Could I go to a field trial and win? Oh no! Why not? After all, the border collie (#1) is a lot smarter than the pointer (#43), according to the intelligence list. Why is it you can't teach the smartest dog in the world to hunt and retrieve quail better than an average-intelligence English pointer? Because the pointer has a brain that is wired differently, which predisposes it to its specific quail-hunting task. It is an interesting difference. The border collie has its pointing behavior wired to its chasing behavior. When the bird flushes, the collie chases it. And that is incorrect behavior for a pointer. Try as I might to discourage the chasing behavior, the border collie has trouble learning the task and it makes this basic mistake over and over again. Thus the hunter has to give the "flush" command and then tell the collie to get down. The pointer, on the other hand, has an innate behavioral conformation such that it will stand fast when the bird is flushed and wait for the shot.

A Chesapeake Bay retriever, a border collie, and an English pointer all have breed-specific behavioral conformations that predispose them to be able to learn their specific task, and to perform that task better than any other breed. The performance has nothing to do with "intelligence" but rather the shape of the behavior. What we have to learn as dog ethologists is to give up the "Aren't they smart!" vocabulary and look at innate behavioral differences.

Thinking about innate behaviors or behavioral conformation is exactly like thinking about physical conformation. The same kinds of genetic processes are at work. It is fairly easy for people to see that the exact size and exact shape of sled dogs are prerequisites for a winning performance. But when we watch a dog herd sheep, we switch gears and start to talk about intelligence and learning. We stop talking like biologists and turn into comparative psychologists. How smart that border collie is! Look how he outmaneuvers those flighty sheep and gets them to go where he wants!

But, when I watch a border collie herd sheep, I'm thinking exactly the same thoughts that I think when I watch a sled dog run. "Smart" is not part of those thoughts. My assumption is that the border collie has a brain that is shaped differently from that of other breeds. The border collie is able to behave in a sheep-herding way for the very same reasons the sled dog can perform in a sled-pulling way. They have a shape that allows them to do their jobs so well that they evoke expressions of admiration from bystanders. But I realize that the border collie probably does not even "know" the sheep are supposed to go into the pen. I couldn't say to the dog, "Go get the sheep and put them in the pen." That is not what is going on in the dog's brain.

What is going on in the dog's brain is part nature and part nurture. But in a very interesting way. Exactly what a dog can learn to do is genetic. Here is where it is important to understand the relationship between the shape of an organ and the limits of how that organ can behave. Intelligence is dependent on how many cells the dog has, and how those cells are wired together. How they get wired together is partly the nurture of the puppy and partly genetically programmed.

This is the underlying assumption of a behavior geneticist. We assume that all dogs have the same number of cells at birth, and we assume that the differences between the breeds are in the way they get wired. If I raise two dogs in the same environment and they behave differently as adults, I assume they were wired differently, and, without other evidence, I assume that is because there are genetic programs that wire these dogs differently. It is not that border collies have genes for herding, but rather, because of gene action, they end up with a differently shaped brain than other breeds.

I keep returning to external shape to illustrate the differences in brain shape because it is easier to illustrate. Let's compare "running" with "intelligence." Perhaps if every time I use "running," you substitute the word "intelligence," the lesson would become clear.

Running is a behavior. Behavior is what selection and evolution are about. Running is what is selected for, not legs. How the animal behaves in its environment is the important point.

Is running genetic? Yes! Are there genes for running? No! Well, not exactly. The "genes" for running and the "genes" for shape (physical conformation) are synonymous.

For example, greyhounds move through time and space differently than do dachshunds. I cannot teach a dachshund to run as fast as a greyhound, because the dachshund is the wrong shape to "learn" to run 36 miles per hour. The genes for running, then, are identical to the genes for the animal's shape.

If I define behavior simply as an animal moving through time and space, then I see clearly that each animal is limited in how it can move through time and space by its shape. A dachshund is limited in how fast it can run, just as the sled dog is. The shape of the running dog is limited by the length of its bones, the angle of its pelvic girdle, the cross-sectional diameter of its muscles, its surface-to-volume ratio. A dog has thousands of different shapes. In fact, a single dog may never have exactly the same shape twice.

But even though there are an infinite number of positions a dog can assume, there are some shapes that the individual cannot assume. A dog is limited in the number of shapes it can assume by the design of its body. A dog can't fly, because a dog does not have wings. A dog cannot think like a human being because it does not have a brain that is shaped like a human brain. A border collie cannot be a good retriever because it does not have a brain shaped like a retriever's.

A behavioral conformation, then, is a detailed description of the shapes that the dog can assume when it is performing its tasks. Earlier, when I described the differences between the *leap-leap* gait of the greyhound, and the *walk-walk* gait of the sled dog, I was describing behavioral manifestations of physical shapes. Just as sled dogs must have an exact size and shape to perform the sled-pulling task well, so must the border collie have an exact size and shape to perform sheep-herding tasks well. At a sheepdog trial, the judge views the working shapes of the border collie herding sheep. If the dog assumes the wrong shape, for instance, grab-bite, it loses points or is disqualified. Similarly, if a show dog stands cow-hocked, the AKC judge, looking at that standing shape, deducts points or disqualifies the animal.

The English pointer and the Chesapeake Bay retriever are examples of dogs whose behavioral profiles are part of their names. Pointers are expected to assume a distinctive, breed-specific standing position: body rigid, head forward, leg pulled up under the chest to indicate that they are on point. Retrievers are expected to fetch and deliver a bird promptly to hand without damaging—or eating—it.

Since these breeds behave differently, I assume (as a behavioral geneticist) that if I look hard enough I will find a shape difference. Shape differences between breeds are often small, subtle, and imperceptible to the unaided eye. Neurophysiologist Cindi Arons looked for these differences in brain shape in several breeds: Siberian huskies, border collies, and livestock-guarding dogs. She found many differences. The three breeds differed in the amount and kinds of neurotransmitters in each brain "organ." The brain is actually a collection of parts, with each part having a specialty in how it moves the animal through space and time. The fact that the breeds had different kinds of neurotransmitters and different quantities of these transmitters is evidence that the breeds' brains are wired differently.

A neurotransmitter is a chemical that facilitates or inhibits the transmission of a signal from one nerve to the next. If the neurotransmitters are different in each breed, then these different brains could not behave in the same way. It is interesting that the brain shapes of those breeds differ in ways you might predict from watching their behavior. Dopamine, for example, is an "excitatory" transmitter and, as you might expect, border collies and sled dogs, the easily excitable breeds, have much higher quantities of this neurotransmitter than do livestock-guarding dogs, which tend not to waste much energy bouncing around. They tend to walk when a collie or husky might trot or run.

The specialized behavior of border collies, pointers, and retrievers depends on where and how much neurotransmitter they have, which in part is a genetic character of the breed. This means their behaviors are hardwired into the brain in some sense. The behavioral display is the result of the genetic shape of the dog's brain. And all that means is that there is an organization of the nerve connections and the neurotransmitters between those connections. Just as the sled dog must be the proper size and shape before it can be trained to run at racing speed, so must the pointer have the pointing-shaped brain before it can show pointing behavior. And if it doesn't have it, no amount of training can get the dog to assume the pointing posture. What a dog is capable of learning is genetic.

Couldn't I modify these hardwired behavioral programs by manipulating critical period events and cleverly employing instrumental conditioning? Yes and no. Could I create an environment or a training system that would modify the greyhound's shape, so it would look and behave like a sled dog? Oh no. That is not possible. In the same sense that the greyhound shape is genetic and difficult to modify, so the border collie behavioral shape is genetic and difficult to modify.

Border collie breeders are so sure that the behavioral profile is genetic that well-bred puppies come with written guarantees that they will clap (show "eye") as adults. Similarly, English pointers will show "point" and Chesapeake Bay retrievers will eagerly search and retrieve, or you get either your money back or a replacement dog.

When I put a livestock-guarding-dog pup out with a farmer, I cannot guarantee it will be trustworthy, attentive, and protective with sheep, because that adult behavior is variable, depending on the environment the dog is raised in. As a breeder I don't have any control over how a buyer raises the future guarding dog, and therefore I can't be sure the adult dog will direct proper behavior toward sheep. However, I should be willing to guarantee that the livestock-guarding dog will not display the border collie clap as an adult. In other words, what is intrinsic behavior for livestock-guarding dogs is that they have no genetic predisposition to show eye-stalk behavior, and cannot learn to perform the clapping motor pattern.

A behavioral conformation should not be more difficult to visualize than a physical conformation. "Eye," "point," and "retrieve" are actually physical characteristics. The dog standing in a show ring has a shape that is being judged. The border collie clapping (showing eye-stalk) in a sheep-herding trial has a shape that is being judged. Pointing at a bird is a shape of a pointer. The show-dog handler directs the shape of the dog for a judge just as the shepherd directs the "eye" shape of a dog toward sheep, and just as the sled dog driver directs the running shape of the team of dogs toward the finish line.

I can't herd sheep with the show shape of a border collie. Sheep won't move away from the show shape of a dog, but they will run from the eye-stalk shape. I can't teach a Chesapeake to assume the retrieve shapes; what I do is direct the retriever toward objects I want it to retrieve.

I picked border collies, pointers, and retrievers for this chapter because each of them displays an unusual and unique set of predatory behaviors. In these breeds, the ancestral forms of the predatory behaviors have been exaggerated, hypertrophied, and ritualized. In each of the breeds a different motor pattern has been modified. What is unique to a breed is the form of the motor pattern and the sequence in which it appears. Thus, if I emphasize and rearrange the motor patterns, each breed becomes behaviorally unique—has its own behavioral conformation. The sequencing of these motor patterns is a product of artificial selection, that is, people adapting the breed to performing its task better than any other breed or species.

The concepts here can be complicated and are frequently misunderstood. The field of dog behavior and behavior genetics in general is accompanied by an ambiguous and abused vocabulary, even among professionals. I shouldn't ever say that border collies have genes for herding, nor should I ever claim to be looking for genes for herding. I never said that sled dogs had genes for pulling a sled.

In Chapter 1, I pointed out how wolves could solve problems more "intelligently" than dogs. By observing their keepers, they could learn to open gates. Some dogs could on some occasions do the same thing, but far less frequently than wolves. Since there is a behavioral difference, there must be a structural difference. There is: wolves have bigger brains. Bigger brain is one character one might expect from animals with more cognitive abilities. But size isn't everything. Chimps have a 550cm^3 brain size as adults. When my very young son passed the 550cm^3

Breed-Typical Motor Patterns

Wild type	Orient>>	eye>>	stalk>>	chase>>	grab-bite>>	kill-bite
LGD	(Orient)	(eye)	(stalk)	(chase)	(grab-bite)	(kill-bite)
Header	**Orient>>**	**EYE>>**	**STALK>>**	**CHASE**	(grab-bite)	(kill-bite)
Heeler	Orient>>	eye	stalk	**CHASE>>**	**GRAB-BITE**	(kill-bite)
Hound	**Orient>>**		**Mark>>**	**CHASE>>**	**GRAB-BITE>>**	**KILL-BITE**
Pointer	**Orient>>**	**EYE**	(stalk)	(chase)	**GRAB-BITE**	(kill-bite)
Retriever	**Orient>>**	eye	stalk	chase	**GRAB-BITE**	(kill-bite)

>> = connected motor patterns   **bold** = hypertrophied   ( ) = fault

Motor patterns of different breeds.

brain volume period, he could tell you the batting averages of all the Red Sox players. Something else besides size is going on.

There are no genes for intelligence. Brains can grow in different directions because of gene action resulting in different shapes. Similarly, there are no genes for pointing behavior in pointers, or genes for eye–stalk behavior in border collies, or for water-loving behavior in Newfoundlands. Nor are there genes for fast behavior in greyhounds or genes for slow in basset hounds. The only way a Newfoundland could become fast is if it could change its size and shape.

This seems to be so simple, but the role of genes is widely misunderstood. Even those who know better—even geneticists—refer to genes for traits: Which breed is most intelligent, or which has the herding gene, or how do herding genes in border collies sort out from the water-loving genes in Newfoundlands when they are back-crossed? These questions make no sense.

Saying that breeds behave differently on a standard test in a standardized environment is very different from saying that there is a genetic difference in intelligence. To say that Newfoundlands like to swim more than other breeds does not mean they have genes for water-loving. If I crossed a greyhound with a basset hound, would I be looking at how the genes for slow sorted from the genes for fast?

I do believe Newfoundlands love to swim in cold water, for genetic reasons, and I do believe Chihuahuas are not a cold-water-loving breed, also for genetic reasons. If they had actual genes for these traits, one could theoretically splice the cold-water-loving gene into Chihuahuas. However, to get that behavior from Chihuahuas, you'd have to make them large, heavy, and hairy like the Newfies. If they look like polar bears they will behave like polar bears. But then they are no longer Chihuahuas. This is what Laurie Corbett was saying about a genetically tamed dingo (Chapter 1).

Loving cold water simply means feeling comfortable in cold water. Or, feeling more comfortable in cold water than somewhere else. The rule from the sled dog chapter is that big dogs have less surface area for their volume than small animals. Newfies are so big, they have tremendous trouble getting rid of heat. One of the clinical problems with Saint Bernards and Newfies is the high incidence of heatstroke problems in year-old animals. After a walk on the first warm day in spring, Bernie often is found pushing his head into some corner trying to alleviate that monstrous headache. My guess is that any animal with the heat load and foot load problems of a Newfie might feel much more comfortable buoyed up and cooled in water, and certainly more comfortable than standing in the Mexican desert like a Chihuahua.

The border collie clapping in front of sheep is perhaps similar. The reason border collies clap in front of sheep must be because that is the position where they are most comfortable—perhaps because it triggers an internally pleasurable sensation. There are many positions assumed by animals that must give them a sense of comfort or pleasure. Certainly the transition from too cold to warm triggers an immense sense of pleasure in me. People are familiar with the pleasure that comes with the different courtship and reproductive motor patterns. The courting animal moves to positions that increase its internal feelings of pleasure. These behaviors are internally motivated and internally rewarded. The animal is internally motivated to search for the environmental signal that releases the behaviors—and those behaviors are their own reward. Thus a dog following the pheromone of a female in heat gains pleasure from the following. The pointer pointing a bird gets so locked into the

Eye-stalk motor pattern by a border collie. A trialer doesn't teach motor patterns. Motor patterns are built into the dog's behavioral repertoire. The handler choreographs a motor pattern ballet, where in the last act the sheep go into a pen.

act that sometimes it's next to impossible to get it off the point and into the next motor pattern in the sequence.

The crouching, clapping shape a border collie assumes is an example of a motor pattern. Or, when a dog flops on its back and urinates in the air in front of another dog, the first dog is showing a submissive motor pattern. Presumably, this is also a pleasurable posture to be in. It's tempting to consider that the dominant position must feel better than the submissive one, but what should be realized is that both are contextually and thus equally rewarding.

In theory all motor patterns are internally motivated and internally rewarded. The animal is internally motivated to search for a mate, and internally rewarded for finding one. Each individual courtship motor pattern is internally rewarded. Copulation could be regarded as a functional behavior where the reward is puppies. But puppies are not the reward for copulating. Feeling an internal pleasure is the reward.

It would be totally ludicrous to think that at the conclusion of the courtship sequence, the dog would need to be externally reinforced with a dog biscuit or a pat on the head—"Good dog!" Similarly, no border collie trialer or sled dog driver ever gives a food reward for performance. The dog already got its rewards by performing the instinctual behavior.

This point is often left out of training books. Most dog-training books are based on behaviorism, the Pavlovian/Skinnerian conditioned response. Behaviorism is based on the assumption that dogs learn tasks because they seek an external reward, or, they are adversively conditioned to avoid unpleasant situations. With the working breeds described here, the dog is internally rewarded and the handler's job is basically to manipulate the dog's location in such a way that the dog anticipates the performance of the pleasurable act.

For any carnivore, eye-stalking, chasing, biting, and killing are similarly their own reward for performance. This is an important point. Does an animal hunt in order to eat? The evidence is just as good that it hunts because hunting is pleasurable. (Maybe animals have to be hungry in order for the predatory patterns to feel pleasurable.) But it just might be that most animals don't "know" that hunting leads to eating.

The dog gets such pleasure out of performing its motor pattern that it keeps looking for places to display it. The animal will search for the releaser of a motor pattern because it gets rewarded so luxuriously for performing. A good retriever sits there and begs you to throw the ball again. (Actually, I've often thought that my dog should be rewarding me with a biscuit for throwing the ball again and again and again.) Why is it so hard to teach some dogs not to chase cars? It is probably not because they are stupid—not high on the intelligence scale—but rather that the internal pleasure released by chasing a moving car is greater than the pain caused by some human screaming *"No!"*

Dogs with hypertrophied motor pattern displays tend to get stuck in a display and are not able to get out. It is the sticky dog syndrome. I had a border collie named Judy who would chase a ball, but as she approached it, she would go back into the eye-stalk instead of continuing on into grab-bite. Since the ball didn't move, she remained in the eye position indefinitely. If I wanted her to wait for me, I would throw a ball into a corner and there she would be when I returned. Sticky dog. I had another collie named Jack who had a nice outrun and good eye, but once in eye, there he remained. "Come up, Jack, *Come up, Jack*!" I would call, but Jack was locked in and didn't want to leave the position, or perhaps he couldn't get another signal through. He was neurologically stuck.

It is clinically common to find individuals of these specialty breeds— the herding dogs or the retrievers—stuck in a repetitive pattern. Our animal-behavior-specialist friends in England, Robin Walker and Peter Neville, think this is a form of epilepsy, and treat it with barbiturates with some success. It is not uncommon in these breeds for a dog not only to get stuck performing the motor pattern over and over, but also to pass out. A border collie I bought in Scotland would slip into unconsciousness in a trial. Sticky is a classic problem with pointers. They get stuck pointing a bird somewhere out there where you can't find them. Every year the pointer magazine has a cartoon of the skeleton of a dog pointing at the skeleton of a bird.

But what is important to understand is that these dogs are not even aware that the goal is to put the sheep in the pen. They don't care if the sheep go in the pen. They just know that the expectation of sheep movement gives them blissful sensation. For the retriever, pointer, or border collie, balls, cars, and virtually any creature or object that has the potential to move becomes the focus of the animal's attention because it puts the animal in a posture that feels good.

Motor pattern is the ethologist's term for an innate or instinctive posture. The *walk, walk* gait of the running sled dog is usually not referred to as an instinctive motor pattern by anyone in the world except me (but it should be). More frequently, the hardwired movements of a border collie clapping are what are commonly called motor patterns.

Individual breeds of dogs can show all or none of the predatory motor patterns. It is rare that they show them all, and also rare for them to show none. One main difference between the breeds is how many predatory motor patterns they show. Another is the intensity with which they display them. Still another is the timing of onset of these patterns (how old is the dog?) and how this affects the incorporation of the motor patterns into the behavioral repertoire.

For example, border collies and pointers begin to show eye-stalk motor patterns at ten to sixteen weeks. The very best livestock-guarding dogs never show any eye-stalk. Yet, even good guardians might show chase and grab-bite, but usually not until they are six months old, and never with the intensity of the border collie. One way to analyze behavioral differences between collies and guarding dogs is to look at which predatory motor patterns are displayed by each dog, and the time of onset during the life of the dog.

With collies and pointers the emergence of the predatory patterns is coincidental with the critical period of social development. This allows them to incorporate some predatory behaviors into their social play. It also results in a social profile very different from that of the guardian dogs, in which the predatory signals don't onset until the critical period for social development is over. Thus, the predatory patterns cannot become part of the social repertoire—which is one reason they can't chase sheep if they have been socialized with them.

The reason I can rear a guarding dog to protect sheep and cannot do the same with a border collie is simply the difference in predatory onsets. I can bond border collies to sheep, but because border collies' social bonding includes eye-stalk predatory components, sheep are sensitive to them and avoid the predatory dogs, which in turn decreases the likelihood that bonding will take place with sheep.

If I see a livestock-guarding dog in a submissive or playful motor pattern (social) in front of sheep I know it is not about to hunt and kill sheep. The dog is displaying dog social behaviors—it is treating sheep as if they were other dogs—and that means social bonding has taken place. If I see a livestock-guarding dog eye-stalking sheep, even if it is part of a social game, I am in trouble because the sheep are programmed to move away from such displays.

There are ways of measuring motor patterns in addition to describing their shape or quality. The frequency of display also indicates a breed's personality. The context in which the motor pattern is displayed is another. For example, a breed is termed aggressive not just for showing growl and bite, but for showing growl and bite at a high frequency, or displaying growl and bite toward humans. All breeds can be aggressive, but some rarely show it, and then only in certain circumstances. Others are aggressive every time someone comes into the yard or house.

A retriever sits in front of me with a ball in its mouth at a higher frequency than other breeds. But, does it sit around the house with the ball in its mouth when I am not there? No! Why not? The behavior has a higher frequency in the presence of people because even though holding a ball in the mouth is homologous with the predatory grab-bite, like the border collie eye-stalk, it appears (onsets) during the critical period for social development. Because the onset is during the critical period, the retriever is only displaying the predatory motor pattern as part of its social repertoire, toward its social companions, which, if the dog is being raised properly, are people. Retrieving becomes a social game played with animals the dog has been imprinted on. This is exactly what I indicated with the walking Pemba hounds, where hunting was really a people-following behavior and not a foraging strategy. Here the behavior is internally motivated, but only in the presence of people.

The pointer displays the point at a high frequency during a hunting sequence, but only in the presence of birds (and people). Now, maybe the point display is intrinsic, but what the dog points at is learned; and in what context it does it, i.e., in the presence of humans, is developmental. The pointer, like the border collie or the retriever, is not hunting (in the sense of a functional behavior), but rather playing social games with the animal(s) it was socialized with.

My student Gail Langeloh raised border collies from wee pups with sheep as if they were livestock-guarding dogs, socializing them with sheep during their critical period. I then took one of the grown dogs and a normally raised sibling to a professional herding-dog trainer in Alabama. I never knew which dog was which. I also didn't tell the trainer about our experiment. This is called a double-blind experiment. He made a fine working dog out of Tick, who turned out to be the normally reared pup. But Fly confused him. "She seems to want to work but she can't hold her 'eye' on the sheep," he complained. We concluded that

Fly had the eye behavior and she could direct it at chickens and balls and even sheep, but something was wrong about her focus on sheep. She couldn't direct a sustained predatory motor pattern toward an animal with which she had been socialized.

Let's add another layer of complication to the story. In a wild animal, motor patterns are usually organized into functional sequences. A functional sequence is a string of motor patterns that results in the completion of a biological need. An example of a hunting, killing, and eating sequence for a general carnivore is:

orient>eye-stalk>chase>grab-bite>kill-bite>dissect>consume.

In this functional sequence, the predator proceeds in an orderly progression through a series of shapes called motor patterns, and each is part of the function of catching, killing, and eating prey. Each shape is an adaptation (the assumption here is of a genetic adaptation, meaning that a neurotransmitter—hardwired—pattern exists) to perform that behavior in the most economical way to get lunch. An animal can't invest more energy in catching the lunch than it gets back from eating it. Five hundred pounds of lion jumps on a mouse, kills the mouse, eats the mouse. Were there enough calories in the mouse to pay back the expense of moving the lion in that predatory sequence? Not only is the predatory sequence the most efficient way to perform hunting behavior, but the size of target to be hunted can also be selected for.

The general carnivore pattern may have been modified over millions of years and each species may have a slightly different sequence, or the shape of individual motor patterns may have been modified—adapted (genetically) to different niches. Some species of predator can substitute one motor pattern for another. Wolves, coyotes, and foxes can substitute a forefoot-stab called a "mouse jump" for the chase motor pattern:

orient>eye-stalk>forefoot-stab>kill-bite>dissect>consume.

Some species of cats have a forefoot paw-clap where they leap into the air and clap an insect between their front paws—a substitute for grab-bite. Other species of cat never use the forefoot-clap because they don't have the wiring even to conceive that this would be a great way to catch grasshoppers:

orient>eye-stalk>chase>paw-clap>dissect>consume.

The canids have several kill-bites. One common variation is the head shake:

orient>eye-stalk>chase>grab-bite>head shake>dissect>consume.

Each of the wild species of cats and dogs has the same general sequence, but the individual motor patterns making up the sequence have different shapes. In cats, with their powerful jaws, the kill-bite is a suffocating or strangling move. The weak-jawed dog family tears flesh, bleeding the prey to death. Both the canids and the felids can direct the kill-bite at the throat, but felids forcefully crush the throat, collapsing the esophagus, while canids move to cut open the throat and bleed the victim. In other words, the behavioral conformation (the shape of the kill-bite) in cats is different from that of dogs.

In some species, one motor pattern is wired together with the next one in the sequence. The first motor pattern releases the second, while the second releases the third, and so on. Some animals cannot perform the grab-bite unless they have chased, and cannot chase unless they have stalked, and so on back up the line. While introducing Namibian ranchers to livestock-guarding dogs, I learned that newborn calves stood a better chance of surviving a cheetah attack than older calves. Why? Newborn calves don't run. The cheetah charges up to the wobbly calf, which just stands there. The cheetah can't perform the paw-slap until it chases the prey. Scaring cheetahs away from their kill is disaster because they find it difficult to display dissect and consume motor patterns without chasing and killing first. The cheetah's relative, the puma, has escaped extinction simply because they "can't" return to carcasses poisoned by predator control agents—they have a behavioral difficulty eating an already dead animal, one they haven't chased and killed.

Pumas puzzled me because they have a caching motor pattern:

orient>eye-stalk>chase>grab-bite>kill>carry>cache,

but they "never" return to the cached carcass. Maybe that was because I was there, watching them, but maybe also because they hadn't performed the initial steps leading to consume. (In captivity, these cats can learn to eat nonfleeing food, but different rules apply in the wild.)

Unlike the cat family, the dog family tends not to have the predatory motor patterns tightly wired together. Canids can begin the predatory sequence starting with any motor pattern, which is why they make such great scavengers. It is rare that any species of wild cat makes its living by scavenging. However, disrupting a wolf in the middle of eye-stalk-chase makes it difficult for the wolf to recycle back to the beginning of the sequence. After all, stalking is an adaptive sneaking-up behavior, and if you are discovered, then the sneaking-up part doesn't work. This is one reason the barking livestock guardian is so effective in deterring eye-stalking wolves.

The display of motor patterns is stereotypic within a species. This means each motor pattern within a species has a species-specific shape. Field biologists can tell—fairly confidently—which predator killed an animal by examining the carcass. Pumas do it one way, coyotes another, dogs a third, and so on. A bobcat kill of a deer is neat because strangling produces no blood, whereas a canid attack is messy. The exact location of the bite is also a good clue. Pumas take a lamb's muzzle in their mouths and suffocate it. But a big puma starts this move and ends up biting a little lamb's face off. In other words, each one of the predatory motor patterns has a genetic basis that varies among species.

What has all this got to do with dogs? Each breed of dog has some or all of these predatory motor patterns. In some breeds, the motor patterns have different shapes, while other breeds may substitute a new motor pattern or delete an old one.

A telling example of this occurred when I first started studying sheepdogs. We had pens with border collies and other pens with livestock-guarding dogs. We fed them stillborn calves from local farms. With the border collies, I could just throw a calf into the pen and they would dissect and consume it. But if I did that with the guardians, the calf would lie there for days. No one would touch it. When I opened the carcass with a knife, the dogs would feed.

Livestock-guarding dogs do not have (and perhaps should not have) the dissect motor pattern. Farmers would call me up and tell stories of how they found a dog guarding a sick and then dead lamb for days. What loyalty, what dedication to the guarding duties! I couldn't tell them the dog wanted to eat the lamb, but didn't have the dissect motor patterns. Then some other farmer would tell me how a coyote tore open a lamb (but the lamb got away) and the dog ate it—live. Awful, disgusting! It is hard to tell farmers that these dogs don't have kill-bite or dissect, but they do have consume. In both cases, I would try to tell the farmers that they had really good dogs, which had shown exactly the correct configuration of the predatory motor patterns for a guardian dog.

The way to create a specialized breed of dog is to rearrange the functional sequence of motor patterns by deleting some and changing the shape of others, and by connecting and disconnecting still others. The basic ancestral pattern is:

orient>eye-stalk>chase>grab-bite>kill-bite>dissect>consume.

In the sheep-herding border collie, it is:

orient>EYE-STALK>CHASE   dissect>consume.

The EYE-STALK>CHASE are in uppercase letters because they are very strong, or hypertrophied, in border collies. They are exaggerated shapes of eye-stalk>chase. Every time I see a border collie eye its quarry and crouch into the stalk position, I laugh, it is so funny-looking. It is almost as if the dog were pantomiming a lion. Grab-bite and kill-bite are faults in border collies (for obvious reasons). But they routinely appear, although weakly and nonfunctionally, and it is rare for a border collie to kill or seriously damage the sheep.

The pointer predatory sequence looks like this:

orient>EYE-stalk>   grab-bite>   consume.

The eye is hypertrophied. Pointers are not supposed to chase, but rather to stay in the eye-stalk position until the handler gives the command to flush (which is a modified grab-bite); but he does not want it to chase the bird. The dog waits and then starts over with the orientation and searching behaviors to find the shot-down bird; then it grab-bites and returns to the handler. Pointers that proceed from grab-bite to kill-bite are said to have a "hard mouth." If the dog then goes on to dissect, of course you flunk it. It is next to impossible to train these two motor patterns out of the functional sequence because they are innately motivated and rewarded. If having the bird in the mouth, which is grab-bite, triggers kill-bite, and if kill-bite triggers dissect, that dog has an incurable problem.

The Chesapeake Bay retriever goes:

ORIENT>   chase>GRAB-BITE>   consume.

Retrievers have hypertrophied searching orientation, which goes almost directly to a hypertrophied grab-bite. Most retrievers don't have well-defined eye-stalk—it's not very useful when sitting in a canoe or a duck blind. Like the pointers, if Chessies follow grab-bite with kill-bite, they are hard-mouthed, a fault. If they have dissect and consume, these are disqualifying motor patterns. We had a delightful Chessie, Scoter, who had her own variation on the Chesapeake sequence: she substituted the motor pattern "pluck" for hard-mouth and dissect, but only with pigeons. Once when she returned a completely plucked but otherwise unmarked pigeon to hand, the judge asked my thirteen-year-old son if the dog could also cook. Tim looked confused because he had never heard of the "cook" motor pattern in dogs.

While I'm here going through the list of breeds, I'll restate the sequences for the basic village dog, the livestock-guarding dog, and the walking hound. The hound goes:

orient>eye-stalk>chase>grab-bite>kill-bite>dissect>**consume**.

But, in the very best livestock guardians it might be:

. . . .**consume**.

Rarely does a good guardian display any of the predatory motor patterns. But here I have to be careful. Just because the frequency of display approaches zero—or because I've never seen it—doesn't mean that it is impossible to find an environment that would elicit the dormant motor pattern. And even if an individual guardian showed some predatory patterns, if it was raised properly with sheep, those behaviors should not be elicited by sheep. In my experience, if guardians show any of these behaviors, then chase and grab-bite occur at the highest frequency. I've only seen eye-stalk a couple of times in the fourteen hundred guardians we've monitored. But I do have a number of reports of perfectly good guardians that hunted and killed wildlife, although I'm not quite sure what "hunted" means. I think the full predatory sequence is extremely uncommon in livestock guardians.

All the predatory motor patterns (except consume) on the breed-typical motor patterns chart (on page 199) can be modified by artificial selection, except consume—all dogs have to eat. With the hunting breeds and the border collie (and perhaps with all dogs), consume is dissociated from the rest of the predatory repertoire. This is interesting, for it is generally assumed that the motivation for performing predatory behaviors is to eat. But for dogs, predatory motor patterns are frequently directed toward nonedible objects such as canvas training bumpers or a thrown ball. In fact, it is sometimes difficult to convert an overtrained dog from training devices to freshly killed game. This suggests that for dogs, the predatory sequence is not functionally motivated—that is, it is actually play behavior. The reward is to perform the hypertrophied motor pattern. One is reminded that sled dogs are also not functionally motivated to run, suggesting again that the motivation is socially facilitated play.

What I've done above is show how a breed profile (breed behavioral conformation) can be changed by taking the ancestral motor patterns and deleting or augmenting one or more. But again there is another layer of complication. A breed's behavioral conformation is slightly more complex than simply deleting and/or augmenting specific motor patterns. Just as in wild creatures, the individual motor pattern can change shape through selective breeding. After deletion or hypertrophy, the remaining motor patterns can be fiddled with. Let's take a closer look at border collies.

THE BORDER COLLIE'S
BEHAVIORAL CONFORMATION

ORIENT   Orient is a searching and an approach pattern. For the border collie, the orient motor pattern is called the outrun. The handler sends his dog in the direction of the sheep. (Similarly, the retriever's handler directs the dog to fallen prey, and the pointer's handler follows the dog as it casts about looking for a bird to point at.) The border collie does not (should not) proceed directly toward the target (the retriever should), but rather travels in an arc, ending up behind the sheep—one hundred and eighty degrees from the shepherd. The shape of the outrun is one of the behavioral conformations being judged.

Some dogs run too wide an arc. I once watched an amateur trial in Scotland where the dog disappeared off the field into the woods at right angles to the line of the sheep. It must have run out around three counties. The crowd was in stitches, just anticipating the dog's return to active duty. The dog indeed did show up, at the far side of the sheep, many, many minutes later. Some dogs have too narrow an outrun. My intense, hypertrophied dog Jane would run straight down the field right through the middle of the flock and then turn around. Sheep would scatter everywhere.

Is there anything I could do about too narrow or too wide an outrun? Yes. The easiest is, get another dog. It is very difficult and frustrating trying to change the shape of the outrun. Oh, sure, I could stand next to a fence or, better, build a fence from me to the sheep in the arc I want the dog to follow. But it is a lot cheaper to buy a dog with the proper outrun. Border collie handlers wouldn't breed a dog with a bad outrun because they think the shape is genetic.

I didn't care that Jane had a badly shaped outrun because I used her primarily as a droving dog or as a predator mimic for testing livestock-guarding dogs. Her outrun was not a problem for my needs. I'd put her behind several thousand sheep on a mountain road where she couldn't get by them. Then I'd play cribbage with the herders and Jane would push the sheep to the top of the mesa, picking up the strays as she eye-stalked-chased (worried) the stragglers. She got that job done just fine. But I couldn't gather sheep with her, and used Flea for that task. If you have a job to do well you must use the proper tool. Each dog has its own behavioral conformation, and good handlers sort among the dogs to find the best one for the job at hand.

EYE-STALK   At the end of the outrun, the dog claps (displays eye-stalk). The clap starts at some specific distance from the sheep. It's almost like a reverse flight distance. Each dog has a specific distance, a threshold distance from the sheep at which it claps.

Clapping consists of several motor patterns. The eye and the stalk can be started from several postures, e.g., standing or lying. The distance from the sheep that these behaviors begin, combined with the intensity of the display, is described as a strong eye or a weak eye. In other words, how hypertrophied the clap is varies among dogs. Handlers prefer a different intensity of the clap in different situations. The intensity of the clap can be selected for.

The speed of the stalk is another variable. Sometimes the dog's stalk is so slow it doesn't move at all. This is called a sticky dog. Or it can be too fast.

As the dog eye-stalks toward the sheep, it reaches another threshold, where it leaves stalk and assumes the chase motor pattern. The distance at which the threshold is crossed is also dog-specific. And dogs all vary as to where the threshold is reached.

This threshold is referred to as the "balance point." In fact, between each of the motor patterns there is a balance point where the dog leaves one motor pattern and assumes the next. What a handler likes about one dog and not another is where the balance points are. Some handlers, in some environments, prefer the dog to be close to the sheep before chasing begins, while others prefer a longer distance. They believe that balance points are genetic and select for the one that pleases them.

These balance points are exactly what I discussed in Chapter 1 with Belyaev's foxes. He thought the balance point between human approach and fox flight was a gene-based character, and he thought the balance point between the distance the fox attained and its seeking cover was also genetic. What Belyaev selected for was not tame per se, but rather he was changing the threshold between the motor patterns.

The good handler knows his dog's threshold points precisely. By commanding the dog to slow or get down, the handler can keep the dog in the desired motor pattern and prevent it from proceeding to the next motor pattern. If the sheep are moving away nicely from the eye-stalk, then the handler doesn't let the dog cross the balance point where it begins to chase. The judge is marking the handler's abilities to know his dog's balance points.

That is what good handling is all about: how well the handler knows his dog's behavioral conformation. If the dog has a bad conformation in one of its motor patterns, then the handler directs the dog in such a way that it doesn't appear. It is the same technique that a show-dog handler uses, in setting the dog's posture in order to try to hide from the judge the dog's cow hocks or some other fault.

CHASE   Chase is a motor pattern and has a shape. The chase in the border collie, like the orientation behavior, is a circling maneuver. The handler controls the shape of the circle by adjusting the speed of the dog. If the sheep are fleeing and the dog is slowed, then the circle looks like a straight line, and as if the dog were following. If the handler speeds up the dog, it goes in an arc around the sheep; and if the speed is maintained, it will just keep circling. Usually the sheep stop when the dog gets ahead, and the dog goes back into clap. Some dogs naturally circle wide and it is difficult to get them back into eye-stalk. Jane, on the other hand, would bust right into the sheep and go from chase to grab-bite if I didn't get her to drop on command at the exact instant she began the grab-bite. Each dog has its own shape of chase, just like the sled dogs have their own particular shape while running.

The handler is like a choreographer who patterns that shape around a flock of sheep by starting and stopping the dog before and after the balance points, as he directs the dog's behavioral conformation in a variety of patterns across the countryside.

Trainability in border collies depends on their ability to be kept on the edge of the thresholds and to be moved from one motor pattern to the next and back again. "Git down!" should drop the dog in midperformance of a motor pattern, to remain motionless and await the next command. I was once doing an exhibition for a television crew and had a remote microphone strapped to my chest. I lost control of Jane, who was in chase mode—she barged right through the middle of the sheep, which were running every which way. I had told her to "git down" in stentorian tones, and I yelled other names at her besides Jane. Then I remembered the microphone. I looked over at the sound man, who had taken off his earphones and was performing an ear-caressing motor pattern. I thought about sending Jane out to Stanley Coren so he could remind her she was #1 on his list of intelligent dogs.

GRAB-BITE   If the sheep don't move away from the clapping dog, the dog goes beyond the threshold point where it should chase, and it can go directly to grab-bite. Oh yes, I lose a wee point if my dog nips the sheep. In the trial ring, the grab-bite is considered a fault. However, many a working shepherd likes to be able to elicit this motor pattern, if indeed the sheep won't move in response to clapping. Sheep get accustomed to being herded, can recognize a weak dog, and just won't respond to being stalked. In this case, the handler has to put the teeth right in their faces. But some dogs cannot do it. The shape of the grab-bite might let them nip at the heels, but at the throat it is beginning to look like kill-bite.

In other breed profiles one might desire a robust grab-bite. For example, the Queensland blue heeler must nip at cow hocks to get these big animals moving. Cattlemen who want a less aggressive droving dog will cross blue heelers with border collies. I had such a crossbreed for a few years, by the name of Bandit. She was a little too soft for cows, which is why I got her, but she turned out to be a disaster with sheep and would injure them with her grab-bite. I relegated her to pet/jogging companion status. But even there she was a problem, because she continually darted in to nip at the heels of the other jogging dogs.

KILL-BITE   Nobody wants border collies to have the full functional predatory sequence. The eye-stalk-chase is as far as the behavior should go. Because the chase is circular, it tends to keep the dog beyond the threshold position where the dog could show either grab-bite or kill-bite.

Still, several trial people have told me that they like dogs that have a little kill in them. What they mean, I suppose, is the latent grab- and kill-bite. There is no question in my mind that in many dogs these biting motor patterns are latent and right below the surface. They don't appear in the trial dog profile simply because good handlers know the balance points and don't allow the dog access to the environmental signal that releases them.

The various arena dogs, such as rat terriers, bulldogs, or pit bulls, also exhibit specific and measurable behavioral profiles that can be assessed in contests. I have no more than a reading knowledge of these dogs, but I have interviewed pit bull handlers, and the conversations about their dogs are very much the same as those of border collie handlers. They have favorite dogs that have some motion they feel is genetic, and they breed for it. There are strains of dogs that are famous because they had some shape of a motor pattern that gave them a unique advantage in the pit. These dogs and their breeders become famous in their field, and

Kill-bite. The dog didn't actually kill the sheep. But what a precise motor pattern it is!

occasionally the dogs become a breed, often named after their breeders: Jack Russell and Bedlington terriers come quickly to mind.

Famous pit bull Whitey Ford showed one of these special motor patterns. As he entered the ring, he would bump the other dog like a sumo wrestler, with his chest. He had a "chest-bump" behavior. He knocked the other dog down, then he'd go after him. Other dog owners liked to breed to Whitey Ford because his offspring had the same move.

My guess is the fighting dogs probably show deletions, substitutions, and hypertrophied behaviors as described for the collies, pointers, and retrievers. The difference is that the deletions, substitutions, and hypertrophy are not in the predatory routines, but in the social-bonding motor patterns, and perhaps the hazard-avoidance behaviors.

I once had a pair of Irish wheaten terriers that did not appear to have any submissive behaviors. A couple of times they fought our pugnacious Chessies, and even though they were losing miserably, they never submitted. From my nap one day I was roused by a gurgling noise and got there just in time to save one of the little tykes. And then he jumped right back in when let go. Luckily, our Chessies could be called out of a dog fight. What happened to the terrier submission reflex? Did artificial selection delete it so these animals would appear fearless (also known as stupid) and therefore able to fight to the death (ultimate stupidity)? On the other hand, I had a pair of Staffordshire terriers that did not recognize or respond to submission in another dog and would continue an attack, even when the other dog had assumed the appeasing posture. But I have made these latter observations quite by accident, and cannot say for sure that these are breed-typical profiles.

There are three subjects we must try to integrate here: the motor pattern necessary to perform the task, the reward system for the performance of the task, and the capacity to learn the task. These subjects are so intertwined, it is difficult to separate them from one another. One risks oversimplifying the performance and the genetics of the behavior.

MOTOR PATTERNS

Innate (genetic), internally motivated behaviors are defined by ethologists as motor patterns. Each motor pattern has an *onset* during the life of the dog. Each motor pattern also has a *rate* of expression. The frequency of maternal behavior, for instance, is episodic during the life of a female dog. And each motor pattern has an *offset* point, after which you don't see it again.

Onset, rate, and offset of behavioral expression vary among breeds and among individual dogs. The timing and intensity of expression can be selected for. The difference between dogs and wolves is the timing of onsets, rates of expression, and offsets of physical and behavioral conformation. Hazard avoidance (fear response), for example, onsets much earlier in wolves than in dogs.

Mother dogs provide a good illustration of the onset, rate, and offset of innate motor patterns with their puppy-retrieving behavior. The mother dog is in her nest. She hears a puppy giving a distress call. She gets out of the nest, goes to the site of the sound, picks up the puppy, then carries it to the nest and places it with the other puppies. (Isn't she smart!) Puppy-retrieval motor pattern onsets after the last puppy in the litter is born. It offsets thirteen days later. After that time, a pup can give the call, but the mother does not respond by retrieving it.

The puppy's retrieval call is a fascinating motor pattern. Each call is identical in pitch, amplitude, and duration. It is a species-specific call just like a birdsong. It is only given by puppies, and only when they are lost. It is given continually until the pup is rescued. The onset of this innate motor pattern is at birth. The offset is roughly four weeks later, at the end of the suckling period.

My Italian Maremmano-Abruzzese, Lina, got caught short one day because of her attention to sheep, and whelped her first puppy in the field. She left the puppy still attached to its placenta and returned to her nest in the barn and had the rest of the litter. The first pup called and called to be retrieved, but Lina was still giving birth. There are some big questions here. Since the puppy was just born, blind and deaf, how did it know it was lost? How did it know which was the "I'm lost" signal? Didn't it need to practice the signal, and didn't it need to be rewarded for giving the signal? No—the signal and the knowledge of when and how to use it are built into the dog; it can't possibly be learned.

Being a good dog ethologist, from my office I recognized the repetitive calling of Lina's lost puppy, and knew why Lina wasn't responding, so I went to retrieve it. I took my tape recorder and recorded the sound, because here was a lost puppy still attached to its umbilical cord giving the correct signal. Fascinating.

Why did I have to rescue the puppy? Well, the puppy had the onset of sending the retrieval call, but Lina didn't have the onset for receiving the signal, because the last pup hadn't been born yet. Six hours later, Lina was ready, but I had already rescued the pup. If the little guy had had the energy to put out the signal repetitively for six hours, I'm sure Lina would have gone back for him, once the retrieval motor pattern onset. But I think the little tyke would have weakened and died long before six hours passed. The call is energetically expensive, and I can't imagine there was enough energy stored in that half-pound puppy.

If Lina was so smart, why didn't she just pick him up right after he was born and take him back to the barn with her? The answer is that if Lina had been a rat, she might have, for the retrieval motor pattern turns on two days before parturition in rats. But Lina is a dog and the retrieval reflex was off until after the last puppy was born.

When my students played the tape recording of the lost Lina puppy to my border collie, Flea, she got out of her nest (none of her puppies was lost), searched for and found the tape recorder, retrieved it, and placed it in the nest with her puppies. It is not lost puppies that are retrieved, it is the object that emits the vocal signal.

At the end of thirteen days the retrieve-puppy behavior turns off in the

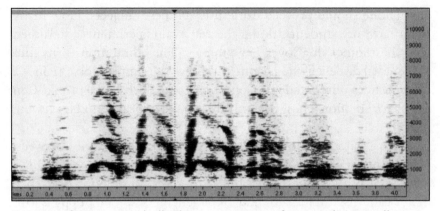

Sonogram of puppy-retrieval call. This is a sonogram of a care-soliciting call (motor pattern). The newborn puppy gives this instinctual call if it is lost. It repeats the exact same call over and over until it is retrieved or becomes too weak to signal. Note the number of harmonics; this is energetically very expensive. Only infant dogs display this motor pattern. Shortly after the neonatal period, the call drops out of the dog's repertoire and does not appear again. (Courtesy of Mark Feinstein)

mother, even though pups will continue to send the signal for a few weeks more. One evening, all dressed up and going out on the town, I heard a weak retrieve signal coming through the pouring rain from our dog yard. Slogging out through the wet, I found the pup, exhausted, out of the box, wet and cold and literally dying. Tilly, the mother, a purebred Siberian (#45, above average on Coren's intelligence scale), was not five feet away, curled up with the rest of the pups in her box. "Tilly," I implored, "what on earth is the matter with you! Can't you hear your pup is in trouble? Don't you hear the retrieve call? Can't you reason, think, recognize, that your puppy is in trouble? Don't you love it?" "Sorry, boss," she said, "but the retrieval motor pattern turned off last week and I don't have any cognitive ability to be aware of the impending danger nor can I deduce it from the information I'm receiving." Heat lamps, sugar solution, and some care saved the pup, and Tilly accepted it when I returned it two hours later.

Scientists are sometimes accused of not being aware that animals have emotions or can think. On the other hand, scientists warn people that they should not be anthropomorphic, giving animals human characters. Well, the retrieval onset and offset lesson should be a good illustration of why one should proceed cautiously on these subjects. How many times have my students, doing the retrieval experiments, exclaimed how the mother dog loves her puppies. The admiration turns into almost total disbelief when Mom can't solve the simple problem once a motor pattern offsets and is no longer part of the behavioral profile. Can dogs have emotions? Yes! But only during the critical period for having them.

There is another motor pattern phenomenon that varies between breeds. Innate motor patterns can differ in their persistence in a dog's behavioral profile. Some, like the "I'm lost" call or puppy-retrieval motor patterns, can remain hidden for generations, if no puppy is lost. All the maternal behaviors for the birthing process are in place at the time of the first puppy, and there is no need to practice or learn them. For example, it is rare that even a new mother doesn't know how to get a pup out of the placenta and chew off the umbilical cord. (Just once I had a female that had the signal wrong and chewed the first pup's ear off instead of the umbilical cord. She quickly made the adjustment and all the pups, including little Van Gogh, were okay.) With intrinsic, genetically hardwired motor patterns there is no learning necessary. The dogs "know" how to perform well, the first time they do it.

But some instinctual motor patterns need to be reinforced or practiced once onset occurs. I discussed this reinforcement earlier, in the chapter on livestock-guarding dogs, but it's important again in this context. Suckling behavior is an excellent example of the need for reinforcement. It is different from the "I'm lost" call of puppies, in an interesting way. Both are in place and ready to go at birth. But after birth, the puppy has literally minutes to "practice" or "learn" nursing. Any puppy prevented from nursing during those critical minutes loses that ability forever. Suckling behavior needs to be reinforced by suckling or it offsets shortly after onset. This is true of most mammals; the results of preventing suckling at onset are often tragic. Human babies who are born with mouth infections or need an operation to correct a palate deformity are thus prevented from nursing until the problem is solved, sometimes days later. They lose the ability to "learn" to suck, and have to be fed mechanically until adult feeding and swallowing motor patterns onset at six months.

Suckling is only one of several emerging motor patterns that need to be reinforced at onset or they drop out of the behavioral conformation. It is true of some predatory behaviors, but only in some breeds of dog.

Livestock-guarding dogs sometimes begin to show the onset of chasing and/or grab-bite behaviors at six months. If farmers are alert and watching their dogs carefully, and remove the dogs from the eliciting stimulus (sheep), then the dogs cannot perform the motor pattern and it drops out of the repertoire, never to appear again. It is hard to say if eye, stalk, and chase would drop out of the border collie repertoire, simply because these dogs self-reinforce the motor patterns in the absence of an external signal. Jane would kick leaves up into the air and eye-stalk them down if there was nothing else to eye-stalk. If the leaves didn't go up into the air, she would imagine they had and eye-stalk them anyway. She was always forefoot-stabbing (pouncing on) imaginary mice.

Thus, breed differences in behavioral conformation reflect not only deletions, substitutions, and hypertrophied predatory motor patterns, *and* the age at which the intrinsic motor patterns onset or offset, but also the persistence into adulthood of practiced and unpracticed motor patterns. Fly, my border collie raised as a livestock-guarding dog, could not hold eye on sheep. Eye hadn't dropped out of her repertoire, but it was modified so much, she was in conflict about showing it to sheep. One might say she learned not to show it to sheep.

What the sheepdog trialer is doing, then, is building a dog. First the dog has to have the right genetics, resulting in the proper, breed-specific motor patterns. Then during development the trainer must encourage the display of these motor patterns in the right environments. Later, with operant and instrumental conditioning, the trainer directs the motor patterns. The working-dog handler directs the dog, placing it in the environment where he wants the dog to display the specific motor pattern appropriate to the task. It is all part of the ballet of sheep, dog, and handler.

The difference from a human ballet is that I cannot teach the dog the correct motor pattern. If the border collie doesn't have the proper behavioral conformation of eye, stalk, or chase, or has the wrong shape for any one of them, it is just about impossible to modify the behavior. And modifying a bad motor pattern is rarely either successful or satisfying. For example, most hunting-dog training books will have some techniques on curing hard mouth. Try this, and then try that, and then . . . get another dog! Top trialers know from experience that dogs with corrected faults, when put under a little pressure in a trial, will often revert to the improper display.

It is very hard for us humans, who have such wonderful learning abilities, to realize the limitations of learning in another species. Why is it that we say wolves are smarter than dogs? Why, if a wolf is so smart, can't I teach it to herd sheep? Why is it a border collie excels at herding sheep, but is not as smart as wolves in solving insightful problems? Why can't the collie be taught to point birds as well as a pointer can? Does smart have anything to do with learning?

Stanley Coren explains very well what the issues are in measuring intelligence. I admire his courage in attempting to rank the breeds according to (working) intelligence, something I would not dare. At the same time I think he does dogs a disservice by doing so. People don't read him carefully and dozens of writers have interpreted the rankings improperly.

I think this is the exact point where a comparative psychologist and an ethologist part ways. I think ranking dogs according to intelligence is something a comparative psychologist would attempt, but not an ethologist. Comparative psychologists look at the dog not in the environment it is adapted to, but rather in artificial, contrived environments. How does this breed compare with another in the "wrong" environment?

Implicit in the claims that breeds can be ranked according to intelligence, or temperament, is the assumption that intelligence and temperament are genetic traits, are breed-specific, and are measurable. It implies that intelligence exists more or less along some continuum. To say that border collies are smarter than Afghan hounds is to assert that the genetic configuration of the border collie's brain enables it to learn more, or more quickly, or perhaps to be able to solve more problems, than an Afghan hound. And it is just not true. A classic experiment that selected for maze-bright (smart) and maze-dull (stupid) rats concluded that within several generations, strains of rats had evolved with different intelligence quotients. Later investigations, however, determined that the rats had actually been selected for their shyness about entering the maze ("stupid"), or willingness to enter ("smart"). The observed results had nothing to do with intelligence.

Certainly Afghan hounds are raised in a different environment than border collies. And we have seen in several instances how the developmental environment affects behavior. Are village dogs more intelligent if they grow up at the dump, or under someone's house, or tied up as pups and fed by a hunter who takes them on long walks in the woods?

How about a village dog that grows up with sheep in the transhumance? Is the temperament of each of these village dogs the same when they become adults? Why is it that highly professional trainers cannot teach a really "intelligent" German shepherd to walk across a grate or not to be shy of a staircase?

Developmental differences change, sometimes radically, the individual dog's abilities, its awareness, its trainability as an adult. My border collies raised with sheep as if they were livestock-guarding dogs could not be taught to herd sheep. Is that because they have less working intelligence than their siblings, which were raised in an environment more normal for border collies? When I raised livestock-guarding dogs with border collies, I ruined them for guarding-dog work. They could not be taught to be attentive, trustworthy, and protective with sheep if they were raised in the incorrect environment. Could I say their working intelligence was diminished by being raised in a home? Is their working intelligence genetic?

It has always impressed—even frightened—me that John Paul Scott and Mary Vesta Marston proposed the critical period hypothesis for dogs in 1950, and fifty years later the dog industry still doesn't seem to understand its significance for traits such as temperament and learning. The dog world seems to be constantly excited by questions of which breed is best, which breed is most intelligent, which breed has the sweetest temperament. From my point of view, the answer to those questions is, the dog that develops normally in the environment the owner is willing to provide is the one that excels. When I buy a dog, my work in shaping that animal has just begun. I want a great set of genes, but the behavior I want is not totally predetermined by those genes.

I believe that thinking about the learning abilities of dogs in terms of intelligence is to miss the essence of dogs. No breed is more or less intelligent in any general sense. They are all just different in what they are capable of learning. The innate motor patterns are the qualities of each dog that the trainer is shaping. And displaying innate motor patterns is what makes the dog feel good.

Innate motor patterns and their internal motivation and reward systems are an important part of a dog's life. The understanding of these motor patterns leads to the ultimate appreciation of a dog's behavior. I see these small pieces of behavior as providing a window into the mind of the dog. To me, the behavioral conformation of a dog is much more beautiful than the physical conformation. When someone says to me at a party, "I have a really beautiful dog and he is so intelligent," I have to change the subject. That person is speaking a language foreign to me.

What I've tried to do in this chapter is show that the behavioral profiles of specialized working breeds are as complicated and distinct as the sled dog's physical profile. The two chapters should generate a synergism that fuses behavior and morphology into a three-dimensional understanding of how a dog works.

PART III

ARE PEOPLE THE DOG'S BEST FRIEND? PARASITISM, AMENSALISM, AND DULOSIS

CHAPTER 7

Household Dogs

WORKING DOGS may enjoy a mutualistic relationship with people. But once evolved into a working shape and behavior, these dogs also get adopted as pets. And here lies a dark side to the association of dogs with people. Just what is the ecological status of people and household dogs? Household dogs—also known as pets or companion animals—are usually considered to be beneficial for people, providing unquestioned love, constant loyalty, eager companionship, and a variety of what people perceive to be positive additions to their lives. But in the scheme of biological survival and perpetuation of the species—any species—what counts is not what we perceive to be beneficial, but what really is beneficial, biologically. For living things, survival depends on a trio of absolutely fundamental needs: food, safety, and reproduction. Unless dogs provide measurable quantities of these essentials, they are not strictly beneficial to the survival of humans.

At the same time, people are usually considered beneficial for dogs, furnishing their food, safety, health care, jobs, and, often, well-arranged opportunities to reproduce themselves. However, as a biologist looking closely at dogs, I become more and more uneasy. At the very least, I think the symbiotic relationship of pet dogs with their owners is seriously unclear. Perhaps that is because there is more than one association between them. Part of the pet-owner relationship is mutual, in the direction of the working dog and its owner. But sometimes I see the population of pet dogs acting like parasites on people. And at other times I see humans treating dogs really badly.

Maybe people don't knowingly treat dogs badly. It could easily be that pet owners are just not biologically good for their pets, sustaining a relationship that ecologists call amensalism.

Amensalism describes the biological relationship in which one species is not affected by the association, but the other species, by accident, is hurt by it. An ecology book describes the relationship between bison and prairie chickens as amensalism. Prairie chickens live in bison country, not really affecting the bison, but as the bison search for food they step on prairie chicken eggs. They don't mean to, but nevertheless they are bad for the birds. Amensalism is usually contrasted with parasitism, in which the parasite purposely lives off the host and saps the host's strength.

Can I cite an example of amensalism? Easily. Take just one case—that of the bulldog. From active and noteworthy employment as the butcher's working dog, and then as a sporting dog, the breed has been adopted as a pet and show dog. In order to enhance its robust, highly unusual appearance, breeders have selected for those traits that emphasize the essence of bulldog—the thick, massive head and short, pug nose. What they have achieved, probably accidentally, are dogs that often can barely breathe, can barely chew, whose puppies are hard to deliver, and females that have to be artificially inseminated. Such animals cannot be living a comfortable life. Their "enhanced" abnormal shape traps them in a genetic dead end. Being caught and bred as household dogs is detrimental to their long-term reproductive survival.

Likewise, I can easily cite an example of the dog being a parasite on humans. Parasites do not usually kill their host, but they do decrease the fitness of the host, and reduce the energy available to the host for its own survival. The two species live together, and one gets its food at the expense of the other. In its simplest form, I go to work, earn a salary, and part of my salary is spent on dog food to feed my pet dog. At its worst, my dog has a disease or an injury that costs me hundreds of dollars to try to cure. Or, it bites someone and I get sued. Or, the dog has behavioral problems and ruins furniture. The dog is a drain on my resources and my energy. The dog takes time and money that I should be investing in my children. It makes me less fit for survival.

The fifty-two million dogs living in households in the United States were probably added to those households as the result of a conscious decision by the householder. People ask me, "What kind of dog should we get for a pet?" What they are really saying is "What kind of dog do you think would benefit us the most?" Often the questioners have been thinking purebred dog. It could be the first time they have purchased a dog, and many people consult friends, relatives, books, and the occasional dog expert for advice.

I often start my answer with "Whatever you decide, get a pup before it is eight weeks old, and spend a lot of time with it during the next eight weeks." I guess my assumption is that if they ask the "what kind of dog" question, they think a breed is a package of behaviors that comes prearranged. They are usually unaware that any and all puppies need an informed and thoughtful owner to shape the pup's course toward well-behaved adulthood. Their assumption is that the right breed is all that is needed to effect the perfect dog-human bond.

They hardly ever ask me, "If you were getting a dog, what would you get?" Sometimes when I suggest breed x, they immediately say, "Oh! But we wanted a big dog." And I ask, "Why would a big dog be more of a benefit?" Are they looking for protection? The conversation suggests that big dogs have more of an aesthetic impact. The big dog enhances their image of themselves. When they say big dog, I just say, "I don't know the answer." I can't fathom a big dog as a companion. I want something that could go with me everywhere in the car. I'd prefer a small dog with a smooth, dry coat, and no long tail to get slammed in the door by accident.

On an estate in central England, I once interviewed Lady Richards, whose dogs enjoy expanses of greensward, intriguing woodland copses, and a peaceful pond. She had a lithe and leggy lurcher (hers was a cross between greyhound and a border collie) that she was very fond of. She said, "We are fortunate in that we can afford to entertain such a dog." She had the right idea: What is it that I as a pet owner possess that will enhance the dog's life?

The relationship should not be a one–way street, where I'll get a dog that pleases me, and if it continues to please me, I won't turn it in to the local shelter.

My dilemma about which symbiotic relationship operates between pets and people is slowly resolving. Especially when I imagine a person purchasing a ten-week-old puppy, locking it in his apartment while he goes to work, coming home to find the apartment trashed, consulting an expensive dog psychologist who after many months says it is hopeless, and turning the dog in to the local shelter, which euthanizes it. Bad situation for the person, disaster for the dog.

Household dogs are what I think of as family pets. I understand there are many variations on the theme. There is even discussion about whether household dogs should be called companions rather than pets. Pets, it is argued, are animals like goldfish or caged birds or snakes. Pets can be exotic or not: the Pet Rock was fashionable a few years ago. Joshua Slocum, on his single-handed sail around the world, had a pet goat for company. But "for company" doesn't necessarily mean companion. I think of a companion as different from a pet. A companion dog accompanies me during some of my daily activities, or shares with me some task, like pretending to be a coyote so I can evaluate the response of a guardian dog.

My border collie Jane was a companion, not a pet. I bought her and her brother as pups from Will Wilson in Scotland in 1977. Jane participated in all our herding-dog experiments, lived in the dorms with my students, and traveled with me everywhere. She was about coyote-sized, and had pricked ears and a sharp muzzle like a coyote. I would use her as a mock predator. I'd send her into a flock of sheep to test the effectiveness of the livestock-guarding dogs. More than once she had to break her outrun and turn tail for me, outracing the guardian dog and at the last minute leaping into my arms, to be cradled and praised. She was also great fun at lectures and demonstrations. It was easy to get her to show the predatory motor patterns. She could not take her eyes off the ball waving around in my hand and the audience would be in stitches. If I put the ball on the floor Jane would eye-stalk it, and if it didn't move she would do a perfect mouse jump on it to get it to move. She joined me on numerous television programs to herd sheep and show the differences in motor patterns between herding and guarding dogs. I used her once in class to demonstrate cognitive differences between children and dogs. From a very young age children understand the pointing finger in ways dogs—even Jane—cannot.

Jane lived in my van, and therefore I spent some part of each day with her. When I was on the road during the years of the livestock-guarding-dog project, we spent twenty-four hours a day together, even sleeping in the van in remote pastures. Although we were constant companions, I didn't really like her. She was an annoying, high-strung dog; she was driven, absolutely driven to display those internally motivated herding-dog motor patterns. She'd stand at the long side window of the van wiping her nose against it as she eye-stalked passing cars. I'd find myself yelling at her to "Git down! Git off my bed!" and milliseconds later she'd be right back. I built her a padded box but she'd start wailing to get out. She never stopped moving. She was always in my face or my ears. It was awful. Would I get a border collie as a pet? Goodness no! It was bad enough having one as a companion.

If a novice dog owner got Jane as a pet, he'd have taken her to Nick Dodman's *(The Dog Who Loved Too Much)* office at the School of Veterinary Medicine at Tufts University for a big dose of tranquilizers. I would have staked my reputation that any of the operant-conditioning specialists would have wound up seeking counseling themselves after a week with Jane. At obedience school, a professional trainer ended up holding Jane at arm's length (while she chewed on it), yelling, "You leave this dog with me, and I'll have her trained by Monday!" I was tempted, but I liked the guy and wouldn't have let him destroy himself attempting to modify Jane into a submissive dog.

Jane illustrates the difference between companion and pet. But I'll call them all household dogs, which is more of an ecological definition. The role of pets, I believe, is the muddiest of relationships between human and dog. I think it is also the least understood in terms of any benefits of the relationship. And, most serious of all, I think the greatest abuses to dogs occur in this home-dwelling population.

Ecologically speaking, as I noted in the Introduction, the domestic dog is an incredibly successful species. Populations of animals grow from small beginnings, colonize the available niche, and at some point reach equilibrium with their environment. As long as the habitat sustains their ability to find food, avoid hazards, and reproduce, dogs do very well.

Right now the population of household dogs in the United States seems to be fairly level. The pet figure may have stabilized because the human population has stabilized, or it may be that people are getting wise and beginning to find out that household dogs are not worth the expense.

Here at the beginning of the twenty-first century, over fifty million household dogs live in the United States. Europe houses an estimated thirty-five million. Dogs in Europe are allowed and routinely appear in restaurants and public buildings, on trains and buses. My experience with friends in Europe may not be typical, because I tend to spend time with people who have dogs. All I know is I have trouble fitting my legs under the restaurant table because of the dogs. As a result, I tend to think the thirty-five million figure is low. If I add Canadian dogs to these populations, I get one hundred million household dogs in the industrial West.

In the United States each year, households produce 3,700,000 puppies. Hobby breeders produce another two million, and half a million are produced by commercial breeders for department store and other retail sales. That is a turnover of 6,200,000 dogs a year. If the population is not going up or down, then 6,200,000 dogs die every year. That is a 12 percent annual mortality rate, which for a species with a life span of a little over ten years is a low mortality rate in the wild.

In the United States, four million of these dogs spend part of a year in animal shelters. For 2,400,000 of them it is the last stop. Almost 5 percent of our companion animals are dogs nobody wants, and they get "put to sleep." Culled. Again, disaster for the individual dog. Some of this culling may be related to competition between people and dogs for food resources. People soon decide they can't afford the dog, and turn them over to humane societies.

What is the biological relationship that fosters the symbiosis between household dogs and people? I'd like to look first at the case for parasitism. If dogs are parasites, and they exist in the United States at the ratio of 1.5 dogs per dog-owning household, then as a human being I am more than a little concerned about this relationship. Among the costs dogs impose on humans are many direct expenses: food, veterinary care, and management (collars, licenses, fences). Dogs also have an impact on society's budgets in the fields of health care, sanitation and medicine, insurance, legislation, and law and order. They act as vectors for disease and as serious nuisances due to incessant barking or destruction of property.

Behavioral ecologists almost always start a research investigation with information on a species' food supply. Dog food is manufactured by companies that purchase the ingredients from the wholesalers of grains and meat products which are also used in human food. They buy on the same auction blocks as do the processors of human foods. Household-dog food ingredients are not leftovers or waste products. The companies will aver that the dog food is not the same quality as human food, that the grains are not good enough to be milled for human food. Some will claim that the animal products are offal and other by-products not edible by people. Sometimes they just say it is surplus human food.

Dog food companies have one big, limiting problem: they cannot frequently or substantially change the formulation of the food. Even tiny changes in ingredients or processing give dogs digestive upsets. A dog eating the same food day after day gets accustomed to that formula. All the little microorganisms living in their digestive tracts are in ecological balance with the incoming ration. Change any of the ratios, and the microbes begin warring with each other, which shows up as diarrhea. This is the reason why changing brands must be done slowly and gradually—so the dog can adapt to the new ration.

Because of this, many people will change brands if their dog's digestion is disordered. Dog food companies are very sensitive to that, not wanting to alienate a loyal customer from his favorite brand. Thus, every day of every week of every month of every year, they have to be buying not only the same ingredients, but the same quality of ingredients. They are not searching around for the lowest-quality waste products. They are locked into obtaining a consistent quality and constant supply. The added cost of buying good raw materials is a minor part of the overall cost of finding the tons of edible materials, and then the manufacturing and distribution of them. Probably the advertising costs more than the ingredients.

Dog food is edible and nutritious for humans, and some people count on it being perfectly safe to eat. (I am not recommending this.) There are reports that in some city districts, more dog food is sold than could possibly be eaten by the estimated dog population. The assumption is that poor people are feeding it to their families. Questions about putting antibiotics or antihelminthics and other medicine in dog foods have to be judged on the probability of its being harmful to humans, because sooner or later some of the dog food ends up in human mouths.

I understand that farmers don't grow food for people, they grow food for money. They grow wheat or beef because it is saleable, not because it is human food. Farming is the process of people turning sunlight into food on farmland. But what is important here is that we in Western cultures have achieved an advanced form of Neolithism, that is, farming with technically advanced tools. All but a tiny fraction of our food is commercially produced by farmers. And our food base is the same as the pet dog's food base. Dog food is not waste products. Therefore its production does have ecological consequence.

Consider this. A dog's normal body temperature is 101.5 degrees Fahrenheit; a human's is 98.6. That three extra degrees means it takes more calories to maintain each pound of dog. In fact, given certain conditions of size and activity, and ambient temperature, dog cells can require as much as twice as many calories as human cells—all other things being equal, which of course they never are. Rising body temperature is not a linear progression. It takes many more calories to raise body temperature from 101 to 102 degrees than it takes to raise it from 96 to 97 degrees. Reports from the Iditarod indicate that a fifty-pound racing dog ingests up to ten thousand calories a day. Several Iditarod champions feed their dogs many times a day during the race in order to avoid the physical loading and digesting problems of feeding ten thousand calories all at once.

Comparing a population of dogs to people would illustrate how expensive dogs are. Assume the average household dog weighs twenty-five pounds, about the size of a small beagle or the Pemba hound (the generic dog). Since many of our most popular breeds are German shepherds and Labrador and golden retrievers, which weigh about seventy pounds, twenty-five pounds seems a low estimate for my illustration. But over six million of these dogs are growing pups less than a year old. So the twenty-five-pound figure might actually be a little high.

I'll assume that the average American weighs one hundred pounds. If dogs need twice as much food per pound as humans, then fifty pounds of dog eats as much as one hundred pounds of human, or two dogs eat as much as one American. Thus, the fifty-two million pet dogs eat as much as twenty-six million people. American dogs eat as much farmer-grown food each day as all the people in New York, Chicago, and Los Angeles combined.

Everybody I tell this to—my friends, veterinarians, biologist colleagues—reacts to that figure as a huge exaggeration. "It can't be!" they say. But I don't think it is much of an exaggeration. It is a tough figure to calculate. The calories needed per pound of dog decrease as the weight of the dog increases. Great, big, reasonably inactive dogs don't require many more calories per pound than a hundred-pound inactive person. But twenty-five-pound inactive dogs require 1.5 times as many. And activity requires more calories. When I add the requirements of growing puppies or children, the calorie needs jump to two to three times the resting figures. Keep the dog outside in the winter and the figures jump again. Even if you wanted to ignore the extra three degrees of a dog's temperature, and consider dogs energetically equivalent to humans, they would still require as much food per day as all the people in New York, Boston, and San Francisco. This for me is still an astronomical figure.

Imagine being responsible for having enough land and growing enough food to feed all the people in three major American cities every day. No wonder the dog food business is such a great idea. Using the back of the bag of my favorite dog food to figure the costs involved, I find that it recommends I feed my twenty-five-pound dog 225 pounds of food a year. At 40 cents a pound, the cost for all the dogs in the United States is in the billions of dollars.

There are other ways of calculating the costs of feeding fifty-two million dogs. I once visited a communal farm in China where the claim was that four hundred people were getting all the food they needed from 150 acres of land. (They were later severely criticized for exaggerating.) That is more than a third of an acre per person, to meet all their nutritional requirements. (In the United States we farm 3.5 acres for every person, but a huge amount gets sold out of the country and we stockpile surpluses, so I don't know how many acres it actually takes to feed us.) The Chinese brigade was growing enough food to feed eight hundred American dogs, or about five dogs per acre. Translate that statistic to feeding American dogs, and it takes fifteen thousand square miles of farmland to feed our fifty-two million pet dogs. (That, by the way, is "only" one percent of our farmland). That is an area as big as Massachusetts, Connecticut, and Rhode Island combined. An area approximately the size of Denmark or Switzerland or Taiwan.

That is also an area four times the size of Yellowstone National Park. Of course, Yellowstone is not prime agricultural land. If we took the fifteen thousand square miles of prime farmland and turned it into national park, we could have the greatest wildlife sanctuary in the world. Imagine the number of wild things and wild habitat that are displaced by the necessity of growing dog food.

So far all we've done in this discussion is feed the household dogs. We haven't considered the pollution from their feces, which has to be equal to or worse than twenty-six million people eliminating al fresco. Neither have I added the medical cost of dog bites, which some experts think are in epidemic proportions. My home state of Massachusetts averages twenty-one thousand dog bites per year that need medical attention. The post office runs a course for employees on how not to get bitten. Society often demands that dangerous dogs be culled. England has dangerous-dog laws, which appear as an almost desperate attempt to protect people from dog bites. And, of course, some few people do get killed by a dog that purposely sets out to kill them. I talked to a man who hated dogs because one pulled his daughter off a bicycle, and when she hit her head she lost an eye.

Some costs are incalculable. How do you compute the cost of lost sleep from the dog barking next door, the cost of running animal shelters, the cost to police forces of finding and returning lost dogs? Most towns have a dog officer who spends all or part of his time on dog problems. Humane societies sponsor conferences for experts who try to establish policy and write intelligent laws about how to protect ourselves from dogs.

Where does all this fit on our symbiotic scale? Is this an example of commensalism or mutualism? The commensal village dogs scrounge the wastes of human foraging. Like rats, they eat what the humans cannot. But they also carry diseases and bark at night. It doesn't matter much to people that they are there, because they exert neither major benefit nor major cost. The mutual working dog performs a service for its food. But household dogs are not only eating at the same table as people, but from the same plate.

Household dogs are competitive for our food and other resources. Their numbers may not have reached the point where this competition seriously collides with our ability to feed ourselves, but that is no reason to ignore the facts. In addition to being well fed, dogs get treated like family members in many other respects. At the same time, they can be in direct competition with human children for the time and resources of the parents. They can inflict substantial harm. If I were a brave ecologist I'd say the household dog sounds like a parasite.

The only argument that would convince an ecologist that household dogs are not parasites would be to show benefits to people for investing the level of resources we do into that population. The benefits would have to outweigh the costs. And I doubt that anybody could do that.

MEASURING THE BENEFIT TO HUMANS
OF THE HOUSEHOLD DOG

The purported benefits of dogs usually fall into two categories: 1) they work for people, providing a tangible service, and 2) they make people feel better in some meaningful way. The category here is household dogs. The subject is pets and companions, not working dogs. The fact that generations ago individuals of a breed were sheepdogs, hunting dogs, or police dogs does not mean that this particular household dog could, even if trained, perform those tasks. Neither does it mean any of those tasks would be useful around the house. In fact, many of the phone calls I get about dogs are from people complaining about an overzealous working dog. And this is absolutely true: while writing this very paragraph, someone called offering me a well-bred Queensland blue heeler— overactive in its breed-typical motor pattern displays, and unbearably obnoxious with its heel-nipping. It was driving its owner nuts.

Among the tangible benefits commonly claimed for household dogs is protectiveness—they are watchdogs. Such dogs supposedly react appropriately to hazards, thereby increasing the hazard-avoidance abilities of the owners. As far as I know there are no data that people with dogs survive more fires, have fewer burglaries, get mugged less, or are more often alerted to other household catastrophes than people without dogs. But the anecdotal evidence always favors the dog. The stories of heroic feats by dogs attract human interest. Cynics call them "Lassie stories." They are stories of dogs that find their way home over hundreds of miles, or perform some insightful act requiring cognitive skills on the order of human intelligence. It is always unpopular to question such miracles. We all want to amplify our relationship with dogs.

But in my experience the average dog can barely find its way home from next door. My cousin Barry's retriever, Rosie, got lost daily chasing cars. She'd start out in front of the house, chase the car to the corner, and become disoriented. I've searched for so many lost dogs I vow each time I do it that I will never do it again. Police blotters have columns of lost dogs and columns of found dogs, and there is seldom a match. Our local radio station has daily lost-dog reports, which they run as a community service.

Dogs don't run into burning buildings, nor are they capable of pulling unconscious people out of one. More likely, the dog knocks a lamp over and starts the fire. The one firsthand experience I ever had with a "heroic" dog was a few months ago when my fishing buddy and I rescued a fellow whose boat had capsized in a fast-running tidal river. He would have been fine clinging to his boat, except his dog panicked and, in trying to scramble up onto the overturned boat, she kept scrambling up her master's back, pushing him underneath the water repeatedly. By the time we got there he was so exhausted he was going down, on the way to being drowned by his own dog. The "heroic" part of the story is that his calls for help were so weak, I'd never have seen him in the dim light except I kept seeing his dog's head come high out of the water, and went to investigate the phenomenon. The dog turned out to be of a breed uncommon in New England, a Catahoula leopard cow-hog dog, which we also rescued.

For me to be cynical about dogs would be incongruous. Of course there are true, unaugmented stories of lifesaving by household dogs. For anyone who has experienced a dog's ability to avert disaster, there is no question whatsoever about the value of the dog in society. And naturally I am well aware of the accomplishments of guide dogs for the blind, of dogs that assist physically challenged people in their daily lives, of dogs used in search-and-rescue and police work. The sight of a well-trained dog performing a task that would be difficult if not impossible for a human is special in the animal world.

In the last two decades of the twentieth century, the status of dogs in relation to people attracted scholarly study and led to the documentation of certain benefits of dogs to the psychological and medical well-being of people. There is much more to this category than buying a dog for a child in order to teach empathy and to care for other living things. I must admit, however, that my motive in acquiring a Chesapeake Bay retriever for our eleven-year-old son was to develop a career opportunity for him. Tim wanted to be a baseball player when he grew up, but since he lacked baseball-playing siblings or friends who would play out of season, he needed dogs to play the outfield. Scoter and my companion dog, Jane, used to retrieve fly balls for him all afternoon. The benefit was not Tim's, perhaps, but mine, since without the dogs I would have had to perform that function. Another benefit for me was that Jane would be so tired at the end that her obnoxious behaviors would subside for a short while.

The psychological benefits dogs can impart to people are measured in both precise and imprecise terms. Psychologists, biologists, and doctors have been measuring vital signs such as blood pressure, documenting longevity, and recording personal perceptions of people in critical circumstances—both with and without a dog. The results in favor of dogs are clear in many cases. Blood pressures in a wide array of social situations, among children as well as adults, are lower. Children are less apprehensive of doctors if there is a dog in the office. People who have a dog to return to after their first heart attack have a better life expectancy than the dogless controls. People with dogs have more access to certain social encounters. It is easier to talk to a stranger who has a dog. Strangers are more likely to talk to you if you have a dog. Walking the dog is a good way to meet other people who are also walking their dog. People confined to wheelchairs find the presence of a dog gains them positive attention from non-wheelchair-bound passersby, which has a positive social effect.

There are good data that say people in wheelchairs with a trained dog feel more like traveling in public. The dog enables them to engage in activities they otherwise would avoid. Blind people often specifically request German shepherd dogs as guides because they feel the dog would confer protection as well as guidance. Even if there isn't a ghost of a chance the dog could cope with a wrongdoer, the mere presence of the "police" dog enables these people to comfortably sally forth. Many people get a specific breed of dog for the home, or to jog with, that they feel would protect them. And if they sleep better because they think the dog is on duty, the dog must be construed as a benefit.

For the behavioral ecologist, however, the problem is to balance the equation. How do you measure the value of these benefits in terms of increasing the fitness of the human population? Old people living longer in a nursing home may perhaps be competing for their grandchildren's funds and their children's time. Spending time taking care of an aging parent past reproductive age at the expense of children at the beginning of theirs may not be to the species' biological benefit.

Again, however the benefit to an individual is construed, it has to be balanced against the costs to the society. The fact that less than one percent of blind people have a guide dog does little to offset the thousands of people each year who get bitten by household dogs.

From the dog's point of view, is this a good relationship to be in with humans?

My own belief is that the dog is not an ecological parasite in the true sense, even if the equation doesn't balance. A parasite evolves to the niche. A parasite benefits from its relationship with its host. Just as a cow is adapted to eating grass, a parasite is adapted to eating its host. Even though the dog may cost the human population, it isn't gaining much biological benefit from its household association with people. In fact, the system people use to propagate dogs may doom those dogs trapped in that system.

Therefore, I think the relationship that fosters the symbiosis between household dogs and people is amensal. Like the prairie chickens whose eggs are stepped on inadvertently by the bison, dogs are stepped on and hurt by people. Although people may believe they are caring for and "improving" their dogs, I'd like to approach their situation from the dog's point of view.

Dogs may appear childlike to some people, releasing caregiving behaviors. It has been postulated that one of the reasons dogs' heads appear more rounded and shorter-nosed than those of wolves is because people selected dogs to look puppylike, which is innately cute to people. Infant qualities also reduce aggressive interactions. But if short, round-ed, puppylike dog faces benefit humans by giving them a release for child-rearing motivations, it is our preference that is being imposed on dogs. It is people who breed the household dog with the short face for their own benefit, without regard for the effect on the dog. Just because the human gets immense pleasure from dispensing care and affection on a dog does not mean that the permanently juvenile dog gets any pleas-ure or benefit from the short muzzle.

In the literature I read on human-animal relationships, the analysis of benefit is almost exclusively in human terms. Dogs make people happy, lower people's heart rates, help people survive heart attacks, increase the quality of life in nursing homes, help disabled and interned people have a better quality of life. Because dogs benefit people does not mean that people benefit dogs, however.

Just because dogs make "ideal" pets does not mean we should make them pets. Just because they have lovely, adaptable social natures does not mean we can exploit them with impunity. And just because they have infinitely malleable shapes does not give us the right to select for any deformation of the basic shape just because a different shape pleases us.

I think dog ownership has reached the point where humans have imposed on the good nature of dogs a little too much. I think that although those interested in dog welfare have done well to focus on the plight of dogs used in biomedical research—the so-called laboratory dogs—and to some extent the working dogs, they are ignoring a far greater and more insidious danger. I think far worse cruelty takes place with purebred dogs, adopted into the pet-class world, and in far greater numbers. I think the household-dog industry has taken too little responsibility for the consequences of present breeding policies and dog ownership.

The production and distribution of household dogs follows fairly standard methods. First of all, household pets are captured animals. "What kind of dog should I get?" says it all. These are not animals that adopt us, as the Pemban dogs do. Household fish in an aquarium are collected from the wild. The parent stock continues to breed in the wild. If the system works well, collectors take only a few "surplus" animals for the pet trade and basically the wild stock continues to reproduce on its own. The beautiful colors and the interesting shapes that we enjoy in our fish tanks are products of natural selection and subject to all the Darwinian rules of survival of the fittest. The stock remaining in the wild stays genetically fit.

It must be remembered, however, that the individual fish removed to aquariums are deprived of their reproductive rights and their genetic potential by our desire to enjoy viewing them. They are effectively reproductively removed from ever leaving offspring—they are genetically dead.

Household dogs are similarly captured. But they come from three different sources. Originally, over the millennia, people captured pups for pets from the (wild) village dogs. Like the aquarium fish, the parental scavenging stock continued to survive and reproduce in its original form. The farmyard dog and my dog Smoky's mother still have reproductive access to a large population of dogs, and are, like the reef fish, giving up a few surplus animals for human benefit. Similarly, capturing a livestock-guarding dog from the "wild" population on an annual transhumance also leaves the wild population intact and with access to a large population of dogs. This situation continuously maintains the genetic health of the individuals. In each of these cases the dogs are still responsible for reproduction, and their normal breeding habits and competition for limited resources will assure healthy offspring.

The second source of household dogs is retired working dogs. Jane was a working dog for me. She got old, and we thought she would be cold in the van at night and brought her in. She became a household dog in her old age. One could imagine hundreds of different scenarios of how a truly good working dog might end up in a home. In fact, the home might just be the most convenient place to keep the dog while it is not working. Many a hunter keeps his bird dog in the house, and on occasion I would bring a sled dog in. There might be advantages to having dogs inside. There's always the possibility of warmth: I've read that Eskimos and Australian Aborigines measure how cold it is by how many dogs it takes to keep them warm overnight.

But the breeding of these dogs is by artificial selection, and perhaps the problem starts here. The modern working-dog breeder breeds dogs that are good performers. I bred Jane just once. She was an awful mother who couldn't tell the difference between her pups and rats. She would eye-stalk them and forefoot-stab them to death if they moved. Here, selection for the eye-stalk behavior had created a biological monster. And I'm afraid that is the risk we run every time we start practicing canine eugenics.

The third source of household dogs is where the problems begin to mount. This is where the householder captures a working dog (breed) for a pet. And then the absolutely worst-case scenario, the breeding of working dogs for the household-pet market. Working dogs should not be pets. Working dogs should not be sold or purchased as pets. And working dogs should not be bred for the pet class of dogs.

If a dog is bred for exaggerated behavioral conformation and is expected to display it in a working environment, it is hard to imagine that the household environment is going to provide the proper stimulation for such displays. Often the person raising the dog has no idea of the critical period of social development or the specific requirements necessary to evoke the proper behavior. This results in dogs that have motor pattern displays not only inappropriate in the household environment, but that also can turn into compulsive disorders. A highly bred working dog raised in a nonworking household environment will still show the working behaviors it has been selected to display, but it will display them abnormally. Worse, it will display those behaviors in bizarre and obnoxious ways. My Queensland blue heeler would nip the heels of joggers. My best lead dog was a border collie that chased cars all day on a suburban street—which was why I got him.

Most working dogs are corrupted, if not ruined, for their job if kept as pets. How can a livestock-guarding dog protect livestock if it is in the house? Many working dogs transfer their socially developed protective behaviors in ways that make them dangerous to people. I avoided passing my guardian dogs on to people who didn't have livestock and wanted them for pets. I succumbed a few times, and I always thought it turned out badly, especially for the dog.

What is the difference between adopting a village dog or a working dog as a pet? The village dog is selected to have a low profile around humans. The village dog tries to maintain contact with the human, which is the source of food. The village dog is selected to solicit care from humans and not to threaten them. Thus, the difference between a village dog that grows up naturally as a scavenger of human resources, and one that is captured as a puppy and raised as a household pet is not all that great. In some cases, as in Pemba, I couldn't tell if a dog was a pet or not, and was totally surprised to find that the vast majority of them were not pets. "Is that your dog?" I asked the girl on Pemba, and it was difficult to discern from the answer what kind of relationship she had with this yard dog. Was it a pet, or was it just a cute scavenger that was protecting her yard from other scavengers? In these societies the difference between casting aside the waste and feeding the dog is a state of mind, an intention, rather than different behaviors.

Adopting an occasional village dog as a pet probably doesn't change the evolutionary dynamics of the wild population very much. The individual dog, of course, loses its place in the wild population and therefore may lose some reproductive opportunities. But it's more complicated for the working dog that is adopted away from its working environment.

The village dog is preadapted behaviorally to be a good pet. The purebred working dog is selected to show a variety of motor patterns that are often obnoxious in a household. Jane wiping her nose on my van window as she eye-stalked passing cars is a perfect example. Her internally motivated patterns (compulsive behavior) are also internally rewarded, and thus practically impossible to extinguish.

The question might be posed as to why one would want a working dog as a pet. Dogs like Jane are limited in their ability to adapt as perfect pets because of their innate behaviors, and thus perhaps they are sought because of the way they look. Something about the way the dog looks benefits the person. Perhaps a sled dog, even though it never pulls a sled, suggests that the owner is associated with the historical legacy or reputation of sled dogs. Perhaps the image evoked by the working dog enhances the owner's status or enables an association with other people who identify with the dog's heritage. Owning a hunting breed suggests that the owner might be a hunter, or might know about hunting and be a self-reliant outdoors person. I often wonder if I had lived in the suburbs whether I would have acquired racing sled dogs. It's hard to imagine they would have been welcome for very long, what with their daily howling and overwhelming cacophony whenever I arrived with the harnesses to hitch up for a training run.

Here is the big shift for dogs: no longer are they chosen for the way they behave, but for the way they look. Selection in the Darwinian sense is for their appearance. The benefit to the human is not in the innate behavioral abilities, but in the coat color, ear carriage, and size. But these are superficial traits, related to survival only cosmetically. It is ironic that the village dog, well suited to survive, is rejected as "just a mutt" for those same traits.

"What kind of dog should I get?" (to enhance my status . . . round out my image . . . amuse me . . .) does not consider the needs of the dog. It is based on the faulty assumption that any breed of dog is adaptable to the household environment.

During the past one hundred years, hobby breeders have taken the working-sporting breeds and bred them specifically for the household market. I understand that throughout history breeders have bred miniature and gargantuan forms of dogs simply for display: the bonsai-garden type of breeding. But few of our modern household breeds are much older than a hundred years. The "perfection" of breeds is coincidental with the interest in expositions in which owners or trainers submit their dogs to judges who decide which dogs are superior in looks. Over the past hundred years, the hobby breeding program has succeeded quite well in isolating subpopulations of working-sporting breeds from their greater populations for the specific purpose of public display and sales to the household market.

This is an important concept to understand. The modern hobby breeder specializes in a breed. A breed is a population of dogs that is mechanically isolated from all other dogs. Only those dogs registered in the breed stud book can be official members of the breed.

When I contrast this policy and process with those of working-dog breeders, who created the breed in the first place, I am horrified at what the hobby breeder has achieved. Breeders of working dogs did not attend dog shows to have their dogs judged on how they looked. They selected the best performers to breed dogs that were good workers. Most of them were not interested in "perfecting" a breed. As a sled dog driver I wanted to win races. What the dog looked like was important only for how that shape could run. Ancestry was important only for what it could indicate about a dog's potential to run.

Actually, of course, I did care how they looked. I wanted them to look like a Coppinger dog. It is a feather in the cap of any breeder of working dogs if his dogs develop a characteristic appearance, along with superior ability, which distinguishes them as being from a particular kennel. Breeds often are named for their breeder, for instance, the Reverend Jack Russell, or Ludwig Dobermann. Hound breeders, especially, strive for distinctive packs. Occasionally, and usually accidentally, a pup with a novel marker appears. A breeder might take a fancy to it and preserve it. The history of the golden retriever, as related in Chapter 3, illustrates this principle very well.

It also points up a very different process than the one producing pet or show dogs. A dog purchased from inbred stock (closed stud book), untested in the field for many generations, is the product of a breeding program (maybe) that has little to do with its working behavior. The expectation of the new owner is that the dog will be good because it is a purebred golden retriever. "What kind of dog shall I get?" "Get a golden retriever because they have a 'friendly nature and disposition, athletic ability, love of water, and natural instinct for hunting and retrieving.'"

What?! That sounds ridiculous to a working-dog person, or to a population geneticist. Friendly disposition is genetic? Love of water is genetic? Athletic ability has something to do with golden color? Is the implication that all goldens have this same set of genes, and all these traits? Is there no variation in golden retrievers? Lord Tweedmouth had good dogs because he had a good breeding program that included a high percentage of crossbreeding and because he hired people to work those dogs from their youngest days and develop the best dogs. He liked to hunt, he liked to have the best hunting dogs, and he was proud of his eye for working dogs. And he culled the bad ones. Anybody who ever created a breed did so by culling the ones they didn't want.

Today's household golden retriever is a caricature of Lord Tweed-mouth's dogs.

The idea that there is something intrinsically desirable in the members of a breed is false. It is the same faulty notion as thinking there is something superior about "royal" blood. It is not only false, but it is bad genetics. It is not only bad genetics, but it dooms any breed that gets caught in that physically isolating trap.

The advantage for pet owners is that in a few generations of selecting for specific size, color, or other superficial reminders of the ancestral working dogs, any of the innate predispositions for the ancestral work almost certainly deteriorate. The rules of genetics say any character that isn't continuously selected for (say, working behavior) will begin to drift because of random events—disease, founder's principles, non-genetic calamities. In fact, what happens in the household world is that owners will quickly start selecting against many of the innate dispositions of a breed because of their obnoxiousness. The innate behaviors will begin to be described as compulsive behavioral disorders. Thus, dogs selected for their superficial traits might eventually make better household dogs because they lose their working dispositions.

But, however, and nevertheless, there is one other genetic mechanism operating. Closing the stud book on a population, in order to promote specific traits, inadvertently and dangerously starts a process of inbreeding. Inbreeding decreases the amount of genetic diversity. Genetic diversity is a source of genetic vitality. This genetic (phenotypic) vitality is good genetic health shown by an animal due to normal actions and interactions of the genes, uncomplicated by any deleterious effects that can be expressed when mutant gene products are paired. The chances of deleterious pairings are increased in inbred animals.

The first rule for all working dogs is they have to have stamina. Stamina is one result of genetic vitality. But stamina plus working behavior make for obnoxious pets. It is rare that any household can tolerate for very long an obsessive dog with stamina. Such a dog is usually much too active to be comfortable in the house. I am certainly not advocating this mechanism for selecting pets, but simply pointing out that sexual isolation and the inbreeding system of producing household dogs are certain over time to produce lethargic, i.e., peaceful dogs. Their personalities are amenable to the needs of their owners, even if their genotypes are in trouble.

There is an even more severe problem occurring in the genetics of the household dog. The working dog was selected to behave in a certain way. The sled dog, for example, was selected to run fast in harness with other dogs. In being selected for that behavior, the dogs evolved a unique shape. This shape allows the dog to behave fast, with stamina. The relationship between shape and behavior is omnipresent.

Therefore, if we want to change the behavior of a dog—make it more peaceful and less vital—we must also change its shape. Herein lies the dilemma for the breeder. The audience wants household dogs that are a historical representation of the working-breed shapes, and at the same time they do not want them to display working-breed behaviors. Trying to select for an acceptable household behavior while holding the working shape constant cannot be done. The dog will come apart. It will show genetic diseases. Its hips won't fit together right. The joints will show weaknesses, and the dog will twitch and bleed and each generation will become increasingly miserable.

Belyaev selected for tame foxes (Chapter 1), i.e., foxes with tame household personalities. And his foxes came apart. He couldn't select for tame and hold the wild-fox shape.

That is exactly what is happening to our household dogs.

Increasingly, the modern household dog becomes a genetic prisoner trapped in an isolated population. With each succeeding generation the behavioral and physical misfits get eliminated from the gene pool while breeders try to hold on to the ancestral form. But in each new generation we see a host of new genetic problems. Lists of breed-specific genetic diseases are now part of the professional and popular literature.

And it is worse than that. Breeders and owners forget what the historical dog looked like. They select for the exaggerated form. They select for the really big ones. They select for the flattest face. They select for the longest face. The breeds end up with weird conformations. Each breed takes on an unnatural shape, becoming a freak of nature. They are loved the way the hunchback Quasimodo was loved—a dichotomy between the grotesque form and the honorable personality.

As the decades go by, every part of the household dog's life is increasingly manipulated for the human host's benefit. The dog is capriciously manipulated for human pleasure. The more bizarre and exaggerated the animal is, the more benefit it seems to confer. This recent breeding fad for the purebred dog is badly out of control. It appears that selection for the exotic is the goal, probably to increase interest and sales. We are producing unhealthy freaks to satisfy human whims. This is terribly unfair to dogs.

The bulldog is perhaps the epitome of what I find alarming and unethical. Take the feisty and useful butchers' dogs of the seventeenth and eighteenth centuries, and turn them into sporting dogs. By the early eighteenth century, determine that tormenting a bull and risking the dog's life just for sport are unethical. The dog then is adopted as a pet-class dog. Over time each of the traits that made it a good cattle-catch dog for the neighborhood butcher gets blown so far out of proportion that the dog now resembles something out of a horror movie.

Their faces are so deformed they can't get their teeth to line up to chew. With some of the miniaturized forms there is no room for their teeth and they fall out upon erupting from the gums. Their faces are so

Wolf biologist Erich Klinghammer and his "pocket wolf."

squashed that the turbinate bones in their nostrils are tiny. Turbinate bones are covered with respiratory epithelial tissue, which helps the dog to breathe and cools its brain. As a result of the tiny turbinates, bulldogs and the other flat-facers have poor brain cooling, poor breathing, and low oxygen tension in their blood. Of the bulldogs that have been tested by the Orthopedic Foundation of America, 70 percent have hip dysplasia. None (zero percent) was reported with excellent hips.

Show-quality bulldogs often can't deliver puppies naturally. In fact, the dogs can't even breed, and have to be artificially inseminated. All of which causes pain. I once asked a woman at a dog show what her Boston terrier was good for. Without hesitation she said, "He is really good at snoring."

Obviously the original bull-baiting dogs didn't have any of these physical characters. And yet someone is always willing to tell the story that the modern bulldog has the turned-up nose so that it might breathe while clinging to the bull's neck. It is fiction. It is rationalizing the purposeful selection for the bizarre. But at the same time I hear people brag how their bulldogs can't even mate anymore. They are amused by these oddities.

The rationalizations for developing freaks are often as bizarre as the traits that result. Breeders propound far-fetched factoids: The komondor has hair hanging down over its eyes so that it can withstand the ultraviolet light bouncing off the snow on winter pastures; shepherds want that forty pounds of corded coat on their komondors to protect against wolf attacks; the great folds of skin on a shar-pei are to prevent serious injury to the underlying muscle in a fight; achondroplasia in dachshunds helps them go into holes after otters or minks; the great bulk of the Saint Bernard is needed to adapt them to tracking lost or hurt people in the Alps.

Of course, nothing could be further from the truth. These are made-up stories. Komondors and the other livestock-guarding dogs rarely have to fight wolves, and borzois and Scottish collies have such weak jaws and ill-formed mouths that they can hardly suck when they are born. There are lots of ways to get dogs into holes other than make achondroplastic dwarfs out of them.

All these traits are genetic deformations found to occur infrequently in all populations. The same genes for achondroplasia found in dachshunds and basset hounds are found in human beings. When the deformation shows up in people, nobody would have the audacity to say these people were selected to go into holes and kill badgers.

People who spend their lives working or hunting with dogs want dogs that look like dogs. There is a basic plan to dogs, and any dog that doesn't have that basic plan shouldn't be bred. The first statement out of any working-dog person's mouth is he wants stamina in his dogs. Certainly a dog with low oxygen tension, or fifty pounds of saggy coat, isn't going to show stamina. I certainly don't want my livestock-guarding dogs with hair hanging all over their eyes so they can't see. Neither do my sled dogs need blue eyes in order to run fast.

It is a terrible joke. Humans take total control of every aspect of the household-dog's life. They are bred capriciously to any shape, size, or color humans think is interesting or aesthetic. Their looks are conversation pieces. A rare breed takes on value for the sake of its novelty. Hobby breeders select for the visual impact of a dog in the show ring. That means amplifying or embellishing an existing trait. Magnify the size, the coat, the nose. Preserve the extra dewclaw, which supposedly helps a Pyrenean mountain dog stay on top of the deep snow in the mountains, but which I always clipped off my day-old husky pups so that as adults they would not tear their skin when they break through a snow crust. I once wrote a book where I designed dog freaks, like flounder hounders, a breed with both eyes on the same side of its head, and then I invented a job for this monster helping fishermen spot fish underwater. If I could figure out how to bioengineer that dog, I could make a lot of money.

A show champion in the household trade is not required to do anything except not bite the judge. Dog shows are comparable to human beauty-queen pageants. Compare each individual with the others in the show and see which one comes closest to some arbitrarily designated, idealistically "perfect" form. At least in the present human version of beauty contests, contestants have to say a few words, or play the clarinet, or recite a poem—to show a behavioral skill.

The difference between these competitions is that the human show has a minor effect on the population's genetics. The dog-show winner, however, becomes the favored breeding dog, and the tiny population that qualifies for inclusion in the breed's stud book is now funneled through these few "best" individuals. To breed heavily to champions is to substantially reduce the effective population size. It channels the available genes through a few individuals. If every owner of a female only bred his dog to this year's champion, then the next generation of dogs would all have the same father. Then the generation after that would all be brother-sister crosses. Thus, the breeding to champions increases the inbreeding coefficient very rapidly. On the surface some popular breeds may appear to have a large population, but their genealogies—their pedigrees—indicate that most of them share close ancestors and therefore also genetic alleles.

The household-dog world is not the only guilty party here. The same problems are now evident in the working-dog world. In the late 1970s I was buying border collies in Scotland, and looked far and wide to find a dog that did not have the champion herder Wiston Cap in its genealogy. (The dogs I found, by the way, were my pseudocompanion, Jane, and her brother Will.) Trialing border collies is a major sport in Scotland, complete with prime-time television coverage of the top trials, and also in the United States, with often generous prize money for the winners. Contestants are eager to succeed, and therefore seek to breed their dogs with the trial winners. Dogs from famous handlers and dogs that win command premium prices for stud services and puppies. But the genetic problems are increasing.

The same reduction in gene diversity takes place when a breed club tries to select against hip dysplasia, retinal atrophy, or some other so-called genetic disease. Every time an animal is culled for a genetic problem, the genetic variation in the closed population is further reduced. It's not just the bad genes that are affected, it is all the animal's genes. Any time there is selection for or against single characters, i.e., "tame" or "hip dysplasia," then one must be prepared for the appearance of new or altered characters because of what Darwin called "the mysterious laws of correlation." Today the phenomenon is called pleiotropy, or saltation— the fact that more than one characteristic can be controlled by a single gene, and selection can result in unintended and unpredictable changes.

When I look at the benefits for the dog in this symbiotic relationship with humans, it looks well-nigh hopeless. Many breeds are living to pay a terrible price for the temporal increase in population or the luxury of expensive food and care. It is not simply that the dogs have access to the kind of medical care that is given to humans, but that they have been bred so they need such care to survive. Breeds like the bulldog are in a dead-end trap. There probably is not enough variation left to get them out of their genetic pickle. Unless the breed clubs open their stud books and allow outside breedings, bulldogs and the other breeds caught in these eugenic breeding practices are headed for extinction. The problem here is that unlike the wild counterpart becoming extinct because of habitat loss, these purebred individuals will increasingly suffer ill health.

What is troublesome is that modern society seems to have little realization of what it is doing to dogs. Owners don't seem to be disturbed about deformation, or even that their dogs are overweight. They are pleased that this is their third Great Dane in ten years and appear proud of the fact that they can cope with its short longevity, giant size, and structural problems. Many, many people love their dogs right into obesity, having no idea of the discomfort from excess heat load caused by the fat.

I believe the modern household dog is bred to satisfy human psychological needs, with little or no consideration of the consequences for the dog. These dogs fill the court-jester model of pet ownership.

From the behavioral ecologist's point of view, I don't know what to call this symbiotic relationship. It may be a new category called reciprocal amensalism. It is bad for humans because of the economic and health problems created by large populations of dogs, and at the same time it is bad for dogs. Individual people may get some psychological benefit, and I suppose there is always a chance for some dog to have a life of true luxury. Nevertheless, for purebred dogs the scope of their existence is to be chosen (or not) according to principles of eugenics—of the worst kind. The breeding programs are not concerned with adapting the dog to the household environment. Rather, the dog is being bred for its showplace value, a not-so-mere-bagatelle of form, with little concern for what's inside, or even if the animal inside that aesthetic shape hurts.

It's a bestial way to treat your best friend.

CHAPTER 8

Assistance Dogs

A FEW YEARS AGO I started to study service or assistance dogs. These are dogs raised and trained usually by an agency to perform some task immediately helpful to humans. These are the guide dogs for the blind, the wheelchair-pulling dogs, the hearing-ear dogs, the epileptic-seizure-alert dogs, and the therapy dogs trained to visit hospitals, homes for the aged, and shut-ins. The category should also include police dogs, as well as dogs trained to find land mines, detect explosives or drugs, and locate people buried in avalanches or under earthquake debris.

There is no question whatsoever that individual dogs have performed important services for people. In some cases, only a dog can perform the specific task. Land mines and especially their trip wires, for example, are now totally plastic and no electronic method of detecting them has yet been found. But trained dogs are good at it. The dogs of the customs services that detect drugs or explosives in huge container shipments save agents a lot of tedious searching. I am always tempted to try to smuggle something by the Boston Beagle Corps, just to see how well those businesslike little dogs can do. I once saw one politely identify a lady who had saved her in-flight apple, and hadn't realized she couldn't bring it into America. Similarly, finding people under snow or rubble is difficult electronically, but trained dogs have abilities in these areas that amaze even trainers of long experience.

Many of the tasks that wheelchair dogs, hearing-ear dogs, or guard dogs perform can be better and more reliably done by gadgets. The guide dogs do provide services that are very tough to duplicate in other ways. However, even in cases where alternatives to a dog can be found, the trained dog companion has been shown to bestow important psychological benefits, enriching the human's life. Often the dog enables the person to go forth into public with a level of confidence and independence inspired by the dog. Because these dogs are regarded as a helpful phenomenon, often their owners receive attention from people who otherwise tend to ignore people with disabilities. The dog provides them with a more normal social environment.

Many stories about service dogs are heroic, right out of the too-hard-to-believe column. But for a person who has been buried alive and owes his life to a dog, any criticism is unacceptable. For someone under stress who has responded positively to the companionship of a trained therapy dog, a critique of the techniques that brought the dog into his life is unwelcome. However, when I looked behind the scenes at the methods of breeding, raising, training, distribution, and use of these dogs, I became uneasy about the well-being of the dogs.

Good assistance dogs are rare. The number of assistance dogs working in the United States is less than 2 percent of the pet/companion-dog population. Perhaps 1 or 2 percent of the blind population now has or would like a dog. The figures for the other service dogs are about the same. The number of customs, police, and military dogs is in the low thousands. The "find you buried alive dogs" are extremely rare, often belonging to hobby clubs or individuals who provide voluntary service in emergencies. Some countries have, as part of their police or military forces, a few dogs they employ worldwide for disaster relief. All the agencies I have consulted with wish they had more dogs. Their common complaint is the lack of dogs suitable for training. Why, if dogs are so adaptable and trainable, are the agencies having trouble finding good ones?

If supply could meet demand, it could well be that in the twenty-first century the service dogs will constitute the major types of working dogs. But in order for this to happen, the industry needs to become a lot smarter about dogs than it is.

Service dogs are interestingly different from any of the other dogs in this book. Mainly, service dogs are mass-produced in assembly-line fashion, by agencies. Over half the dogs that enter the system fail. The failure rate is due in part to abnormally high levels of "genetic" disease, to the inability of many dogs to respond properly to classical or instrumental conditioning, and to high stress levels created by rearing and training programs.

Throughout this book my job is to look at the evolution of working-dog behavior. For each type of dog, I have described the adaptive processes that led to the symbiotic relationship between dog and person; and also, which symbiotic relationship works for each type. In this chapter, I have to ask the same question: Is this purportedly euphoric relationship of biological benefit for both people and dogs?

In the case of household dogs, trying to define which symbiotic category expresses the true relationship between them and people is an infinite debate, partly because of the variations within the population of dogs and the populations of people. But I don't think that we are going to have any problem determining which symbiosis exists between service dogs and people.

On the surface, it appears to be true mutualism. People believe they are providing dogs with satisfying work and good care, and "obviously" the dogs are enhancing the lives of physically challenged people. Actually, the symbiosis existing here is called by some behavioral ecologists "dulosis," or slave-making. The term is usually applied to species of ants that capture other organisms and force them to work for them, causing the loss of biological benefits for the captured organisms.

There is a sticky point here. Some anthropologists feel that all domestication is an example of dulosis. The argument is that all the domestic species have been "captured" and genetically and behaviorally transformed. Domestic animals are manipulated for human benefit. The argument between welfare groups and others who depend on animals for their livelihood, such as the agricultural community, is whether animals have only instrumental value or whether they have intrinsic value, apart from their usefulness to humans. I have been to meetings where I have heard discussions about all the domestic animals being slaves (instrumental value), created (eugenics) to be eaten, made into shoes or sports jackets, forced to help us wage war, as in the case of the horse and dog, or work as beasts of burden—whatever we human captors please.

The argument in favor of domestic animals rests heavily on the assumption that people created them for their own benefit by artificial selection. They are a creation of people for the benefit of people. The rationale for keeping them is that people domesticated wolves to pull sleds, hunt, or be companions for human beings.

However, this book makes a different argument. Dogs evolved

through natural selection to scavenge from people. Through environmental circumstance some have evolved from the original commensal conformation to the shapes of hunting dogs or livestock-guarding dogs, with little direct intervention by people. Increasingly over the centuries, the working dogs get captured and put to work, where the best of them survive genetically. Then the household dogs get captured from the working-dog populations, where their survival depends on the capricious conception of a historical representation of a conformation. Some of these new breeds get trapped in genetic isolation, a dead end that precludes natural adaptation to the captive environment. The pet/companion dogs, bred for exotic form, are almost the peak example of how the evolving symbiotic relationship goes badly for dogs. Almost!

The service dogs are perhaps the most exaggerated case of eugenic manipulation of an organism. There is little question that the service dogs represent a great benefit for a very few humans, but are also a biological disaster for any dog that gets trapped in the system.

Why aren't sled dogs, herding dogs, and gun dogs also examples of slavery? I believe that a case to that effect could be made. But then, I would divide dulosis into two subsets.

One: The forced laborer has breeding opportunities based on superior performance. In these cases artificial selection mimics natural selection, where the best-performing animal leaves its genes to the next generation. Performing animals are well fed and well cared for. The human tends the animal in such a way as to provide the opportunity for the demonstration of superior performance. Superior performance imparts to the animal a value by humans for their breeding potential. Animals can retain this value after their performing years are over. Over time, selection favors animals that are predisposed to the task and appear to enjoy it. In fact, it is very hard to stop such dogs from spontaneously going to work. Their adaptation for the work is reflected in their natural abilities and eagerness to perform.

Two: The humanitarian-service dogs are animals captured from working-dog breeds, or more likely from household breeds, and prepared for work by an institution that specializes in preparing dogs for specific tasks. To enhance their continuous benefit to people, they are sterilized. The individuals' healthy sexual organs are amputated so that, like human castrati or harem eunuchs, they obtain a secure and prosperous life as long as noble service is performed.

For the vast majority of the world's four hundred million dogs, their relationship with humans tends to enhance their biological survival. For captured household breeds and service dogs, genetic survival becomes questionable. The genetic trap that the household dogs find themselves in can sometimes be overcome by escaping back into the larger population and reappearing as mongrels. Their genes can stray back into the feral village-dog populations. Village dogs can be re-created over and over again, adapting rapidly to local or changing conditions.

But there is no hope of escaping for any dog that gets captured by the service-dog industry. The seed of the few dogs retained for breeding produce sterile workers, so rendered by agency policy. Superior performance has no biological reward.

These dogs are supported by fund-raising activities and advertising. The owner/producers are specialists in managing and marketing, but don't know very much about dogs. They hire trainers and veterinarians, who process puppies for service. The brochures feature individually excellent dogs, and graduation ceremonies of human-dog teams. It is standard business and product sales technique. Sell the idea to people donating money. Calendars, newsletters, and promotional materials extol the wonderful bond between humans and the agency-trained dogs.

It is all about people. How dogs help people. It is about how dogs enable people to go out in public. It is how this person's life has changed because of the dog.

I go to conferences, and for me they are boring. No offense, but I like studying dogs, and at these conferences rarely does anyone mention dogs or dog behavior. It is all about how much better people feel with an agency-trained dog. It is about how to get insurance to pay for the agency-trained dog, and who should regulate activities, set standards, get the grants. Sometimes there are a few service dogs attending the national conferences. Often it looks like the dogs don't know where they are or whom they are with. They look as bored as I am, sitting through uninteresting sessions that are not about dogs.

To me the life of a service animal does not look secure, comfortable, or noble, and certainly not euphoric. I see many wheelchair dogs and guide dogs that don't look at all happy with their lot. They look like animals that are results of aversive-conditioning techniques. These are animals taught to avoid problems. They walk down streets watching for some terrible event to happen. Don't run, don't chase, no sudden moves. They are picked because they have no innate tendency to chase a cat or squirrel or fight with another dog. They are wimps avoiding getting into trouble. They don't know why they are pulling a wheelchair, or what happens when they push a light switch. They show little to no evidence of internal motivation to perform these tasks. They are animals that have little comprehension of what they are accomplishing. And they are caught, with no chance of getting out except to flunk out. It is a stressful way of life.

The major guide-dog agencies raise their own puppies. The biggest wheelchair-dog outfits raise their own dogs, although the females are farmed out to private homes for whelping. Many service-dog agencies and especially those that train hearing-ear dogs use dogs from shelters. The customs service uses almost entirely shelter dogs, but some are donated and some are purchased. The therapy dogs are almost all rescued dogs, and many of them, in some agencies, are rescued greyhounds. The search-and-rescue dogs, police dogs, and bomb-sniffing dogs are often bought from breeders. Many are German shepherds produced specifically for, or by, the agency that raises and trains them. There are exceptions in each of these categories, and an individual agency, of course, can change the procedure over time.

Whatever system the agencies use for acquiring dogs—be they animals reared in-house, or purchased or donated puppies, or rescued dogs from shelters—the majority fail. A 50 percent flunk-out rate is the number just about all the agencies admit to. Trying to get accurate figures is difficult. Publishing failure rates is not good publicity, and even honest answers can be favorably colored. Questions are often answered with "It depends on how you count," which is fair enough and true. The fact is that no matter how you count, the percentage of dogs that graduate is not great.

The failure rates vary. The consequences of the dog making a mistake also vary. A therapy dog doesn't need to be much more than trustworthy and polite. The hearing-ear dog failing to signal the ringing telephone or the wheelchair dog that can't press an elevator button can still be a good and safe companion. If the dog guiding a blind person steps off the curb into traffic or the mine-detection dog doesn't concentrate, the results can be serious.

Why do dogs fail to graduate from the institution? One outfit told us that 65 percent of the failures are related to health. Imagine, purposely producing a stock of dogs where 30 percent (50 percent overall failure rate times 65 percent health problems) are unhealthy. Other agencies report health problems as low as 14 percent, but behavior problems account for 96 percent of the dropouts.

The major topic of discussion surrounding service dogs, year after year, is the issue of genetic health. (Here I am, fascinated by behavior, sitting at these conferences through two and a half days of genetic health problems in order to get to half a day of behavior.) Skin diseases are a growing problem. Hip dysplasia has been a problem for forty years. Elbow problems and general arthritis are constant and nagging problems across the industry. Progressive retinal atrophy is extensive in several breeds, and ironically a huge problem in guide dogs. Million-dollar research projects are trying to identify genes for retinal atrophy. The reasoning goes: if you can find genes associated with retinal atrophy, then you can screen your breeding stock and you can identify affected individuals as puppies and eliminate them from the rearing and training program. Some agencies don't start training their puppies until they are a year and a half old, simply because that is the earliest age when some genetic diseases can be detected. Whether a year and a half is the correct age to start training a dog is not an issue. This age is set simply because it is foolish to invest any training time into an animal that is going to develop a disability.

Does that policy conform at all to the developmental biology of a dog?

Name an organ system—blood, bone, muscle, liver, adrenal, kidney, skin, eyes, ears, reproduction—and there is more than one agency with dogs debilitated in at least one organ. And yet the agencies keep breeding more dogs each year. I asked a chairman of one agency how they were trying to increase the number of good dogs, what with the increasing failure rate, and the straightforward answer was "Produce more dogs."

Some people might think it unethical to breed animals that are carriers of genetic disease. If the disease rate was 10 percent, they would cringe at the dogs' prospects. Certainly if human parents knew they had a 10 percent chance of producing a genetically defective child, they might seek counseling and might decide not to take the chance. But the dog industry tells me that a 15 percent incidence of hip dysplasia is low, even a good rate. I guess the reason they think that is because some service agencies find as many as 35 percent of their dogs with genetic diseases.

Welfare societies, whose job it is to monitor the well-being of domestic animals, are somewhat schizophrenic on these issues. Recently the Iditarod came under the fire of welfarists. The history of the race, run since 1973, shows that out of 20,000 dogs entered, 145 died while racing—a figure of seven-tenths of one percent. This is bad? To me, this is a sign of a really healthy population of dogs. It compares highly favorably with the 35 percent of assistance-dog populations with lifetimes of debilitating genetic anomalies.

Many service dogs flunk because of behavior problems. In some agencies the flunk-out rate due to behavior problems alone is 40 percent. Leading the list are aggression toward humans or other animals, shyness, and fear. In addition, a service dog must have a temperament that is not so robust that a disabled person cannot control the animal. At the same time, the dog must not be so timid that it could be forced unknowingly to place the person in a dangerous situation. The dog must have the strength to disobey an order if it would place the client in jeopardy. Dogs that aid people should not be tempted to chase cats or squirrels, or to socialize or fight with other dogs. They should not be overly friendly with people or seek attention. In other words, they must have quite special, almost undoglike dispositions.

A service dog should be the correct size. Labrador and golden retrievers are the breeds of choice for wheelchair dogs because they are the right size, rather than because they have the ideal behavior. The guide dog must be large enough to carry the harness properly, but not so large that it outpaces its client. Rarely do you see a Chihuahua or a dachshund as a guide dog. Just as with the sled dogs, each dog has a "reach" that determines its speed (of walking), and each person has a "reach," which changes with size, age, and physical ability. Thus service dogs have to be carefully matched, according to size and gait, with the recipient. One agency has bred reasonably small German shepherds because people returning for a second or third dog want a shepherd ("just like their first dog"), but are now older and don't have the stride or ability to control a big dog. Hearing-ear dogs are often smallish, feisty little dogs that jump up on the person as a signal. They must not be so big they knock the person down in the process of signaling.

Each dog has to be perfect for the person it goes home with. This matching often requires careful attention to details; a dog that is timid about crossing sidewalk grates, for instance, must be matched with a person who lives in a grateless environment.

I believe behavioral failure is built into many of these service dogs during the rearing process, which is based on convenience and cost. Yet for the livestock-guarding dogs and the hounds, and indeed all the dogs in this book, growing up means growing up in the correct environment. Throughout the service-dog industry there appears to be major ignorance about early development and its effect on the production of good dogs. There is also a belief that any dog will do. Any good, well-bred dog should respond to classical or instrumental conditioning. And if the dog does not respond well to standard training procedures, it is often characterized as not well bred, or having bad genes. If the animal is aggressive to people, or timid and fearful, there is a tendency to think it is the nature of the dog, that the behavior is inherited. Therefore, goes the reasoning, more rigorous selection must be followed to avoid the behavior.

At the same time, trainers seem to know that not every dog responds to Pavlovian and Skinnerian conditioning in the same way. Breeds of dogs respond differently. Labrador retrievers are more likely to be timid and shepherds more likely to be aggressive. And that is a genetic difference—isn't it?

There is a technical literature about breeding timid and aggressive traits out of guide dogs, most of which comes from Australia. If you can breed timid out of a strain of dogs, then it must be genetic—right?

The important point that most agencies don't seem to understand is, phenotypes are a reaction to the environment they develop in. Grow the same set of genes in three different environments and you get three different results. By keeping a dog away from people during its critical period, I can create a dog that is "genetically" (and permanently) shy of people. Can I breed this trait out of the strain? That's not necessary.

If some behavior is genetic, would you expect breeds of dogs to be different? Absolutely! Look at Chapter 6 on herding dogs. All those breeds were selected to have different onsets and frequencies of innate (genetic) behaviors. If I have two breeds (different sets of alleles) and I want them both to perform well as guide dogs, do I raise both breeds in an identical environment? Absolutely not. I would think it is pointless to raise a retriever and a border collie in the same environment if I wanted them to become guide dogs. In fact I think it is pointless to train a trial-quality border collie or retriever to be a guide dog. Neither could respond to behavior modification, with their particular behavioral conformations, in such a way that they could be good guide dogs.

There are subtle problems that lead to a dog's failing that are not particularly the dog's fault. For example, dog breeds develop reputations, and those reputations color people's interactions with them. German shepherds often fail because they are seen as people-aggressive; so are pit bulls, rottweilers, and Dobermans. Labs and goldens are seen as friendly breeds. In films, I've never seen a golden retriever as an attack dog used by some sadistic criminal to tear up his runaway girlfriend, or to keep prisoners in line. But I have seen those scenes with Dobermans and German shepherds. In fact, when I ask some agencies why they persist in raising German shepherds, they tell me that some of their clients feel safer with shepherds, meaning that the dogs not only guide them but protect them. It is a cultural assumption that shepherds are more aggressive to potentially bad people.

The fearful responses of people to a perceived aggressive breed "teaches" the shepherds or pit bulls to be aggressive with people. As the dog walks the streets, some people, almost imperceptibly, will take a step back or away from the dog. In two weeks the dog can become aggressive toward people. If people treated a golden retriever the same way, in theory one would get the same results.

Are shepherds genetically aggressive? Yes! Where are the genes for aggression? In their coat color and shape. It is a feedback system, where each time a person steps back from the shepherd because of its coloring and shape, the dog becomes more responsive to the move, and the people react more demonstratively to its movement, and so on. Can you train the dog not to be aggressive once it has learned to be? Probably not satisfactorily.

Okay, then can you *breed* people-aggressiveness out of shepherds? Of course! I'd start by breeding shepherds to have yellow coats and floppy ears. Is the only genetic difference between German shepherds and yellow Labs that they look like German shepherds or yellow Labs? I don't know. But I wouldn't waste my time trying to breed aggression out of shepherds without finding out where the aggression is coming from.

The main point of all these controversies is that we do not have good ethological definitions of behaviors such as aggression. Is aggression a unitary behavior, starting in one center in the brain? Or are aggressive behaviors specific to the contexts in which they occur, such as foraging, food-pan aggression, reproductive aggression, competition hierarchies, hazard-avoidance aggression, fear-biting, and so on, and therefore have different developmental trajectories? Without knowing which specific aggressions we are working with, it is hard to control them. A sled dog driver doesn't want his team dogs to fight, so he avoids food-pan dominance routines by not feeding them in the same pan when they are growing up.

I think the service-dog industry promotes poor behaviors precisely because industry-produced dogs have chopped-up lives and are products of disconnected events. Each stage of development is somewhat traumatic and does little to prepare the dog for what is coming next. Not many of their critical period activities are designed to prepare the dog for its adult tasks. The expectation is that a couple months of training, beginning at a year and a half, can make any dog a good service dog.

Often puppies are born into sterilizable kennels, where they commonly spend up to ten weeks, over half of their critical period of social development. Kennels at the very best agencies are an impoverished environment. However, the agency mission for the puppyhood period is to keep them from dying of puppy diseases. The prevailing attitude is that the only purpose of puppyhood is to develop immune responses and mature to training age. The purpose of kennel workers and caregivers is to keep the pups clean, well fed, and disease free. These caregivers wear lab coats and uniforms, or even the surgical variety of moonsuits. For a pup, it is like being born into a well-run, efficient orphanage. The animals are physically well cared for, but at the same time little is done for their emotional needs. Is anyone preparing them to climb stairs or cross grates? Is anyone thinking about olfactory imprinting? Is anyone preparing them not to be anxious about traffic noises? A good gun-dog person knows pups must hear guns exploding during the critical period, to reduce the chances of their being gun shy as adults. Why is it so many service dogs flunk because they are shy of traffic, cars, or horns? It is because they had no exposure until they were ten or more weeks old. Or it is because their first exposure to the outside world at ten weeks was to be taken from the sensory-impoverished kennel on a traumatic car ride.

During the critical period of socialization, from birth to about twelve weeks, these animals see only vertical, standing-up people, who talk and move normally. They play with other puppies. And yet they may have to spend their adult lives with a person confined to a chair, with no other dogs around. Blind people and wheelchair-bound people move, and often speak, differently from other people. The socializing puppy must be exposed to the people with whom it will eventually work. The puppy, which learns with its five senses what its normal environment is, must be taught that the normal environment contains crowds, highways, or persons in wheelchairs.

My experience is that the people who manage kennels really like dogs, and even love the puppies. However, it is the puppies—as puppies—that are the focus of their attention and care. Constraints of time, money, and facilities govern agency policies, and usually mean that a dog's infant and adolescent environments have little to do with its adult life. Many agency managers have never trained a service dog. Very few of the staffs in the United States are familiar with the literature available about puppy socialization, and even fewer take advantage of it in their puppy-raising policies.

In every other working- or sporting-dog group there are continuous expectations for each puppy, and early experience and training programs are carefully tailored to what that pup will be doing as an adult.

What the service-dog industry does do is attempt to develop a general system for puppy testing. The goal is to develop tests for pups, perhaps at various ages that will reveal what the dogs will be like as adults. To me, this is a bizarre concept. Among the underlying assumptions is the notion that the behavior of the adult dog is predetermined by eight or twelve or twenty-four weeks of age. But adult behaviors are not part of a dog from birth. Behavior is always a response of the genetic program to the environment. Yet the idea of testing pups persists, as if the early discovery of timidity in a pup would ascertain timidity as an adult, and therefore that pup needs to be discarded before time and money are invested in its first year. This policy totally ignores what is known about dog development and behavior.

There does not seem to be much institutional realization that in every other field of working dogs, the good ones are carefully developed from tiny puppies. If I were offered the best-bred sled dog, or hound, or livestock-guarding dog that was raised in a kennel for the first eight weeks, I would refuse it. The chances that such a dog would turn out well get slimmer and slimmer with each day that goes by. I have asked the personnel at several agencies if they had read the classic Scott and Fuller work on critical period for social development. One geneticist had "heard of it." That would be like me saying, "Oh, yes, I have heard of Darwin." End of conversation.

In the decades immediately after World War II, great progress was made in understanding dog behavior and socialization techniques. The theoretical work of Scott and Fuller at the Jackson Laboratory in Maine was put to practical use by Clarence Pfaffenberger and his associates at Guide Dogs for the Blind in California. Almost by accident, they discovered that pups raised in homes from the age of twelve weeks made better guide dogs than those kept in kennels until they were nine months old. Further research in California led Pfaffenberger to conclude that great improvements could be made in both temperament and structure by what he called "intelligent" breeding programs. Toward the end of his 1963 book, *The New Knowledge of Dog Behavior,* he was reporting no cases of structural lameness in the previous two years, and painting a rosy picture of the future of guide dogs due to research. But now, nearly four decades later, this guide-dog agency notes just as many failures because of physical and behavioral problems as do the others.

What is the problem? Why are Scott and Fuller ignored? Several outfits have hired modern geneticists who apply mathematical formulas to biological problems, trying to select against disease and for good behavior, but without paying attention to the dog as dog, the puppy as developing puppy. They are treating the dogs as if they were genetic robots, as if all features could be genetically controlled. There are no genes for hip dysplasia or behavior. Both are epigenetic.

After the initial kenneling, puppies go to a nice home for up to a year and a half. The home can be, more or less, another waiting process before the real work begins. Some agencies ask puppy walkers to do some training and pay attention to basic behavior development. Some agencies try to head off problems by observing the dogs on site once or twice a month, or bringing the dogs back to headquarters for evaluation and training.

Most agencies do not. The biggest agencies are often the worst offenders. An agency raising six hundred or more puppies a year might have puppy walkers scattered all over the Northeast or the Southwest. Just the sheer distances between individuals of this year's crop make it impossible to check on most dogs. Essentially, the puppy walkers raise the dogs, and the agency waits to see if the dogs develop one of the "genetic" diseases or behavior problems. This process of waiting can take a year and a half.

If the dogs pass the waiting period, they then go back to the agency's kennel. They are inspected and caged. They have to learn to defecate on the floor, and they also learn to participate in raucous barking. They are given a week to adjust, after which they join a string of dogs managed by a handler they have never seen before; the handler has eight new dogs he has never seen before. In some agencies, dogs are out of their cages for as little as twenty minutes a day, when they go through their daily training routine. None gets more than an hour. In other agencies, dogs are trained in town right from the start, and thus are out of the kennel longer than the formal training period. Still others are trained in the corridor off the kennel. For some home-raised dogs, the shock of being dumped in a kennel cell at a year and a half old for four months of training is too much. I see "dropped because of kennel stress" written on their report cards.

The trainers are a mixed lot. Some are very, very good, with years of experience. They like dogs and they like training dogs. The combination of experience and love of dogs makes a true "dog person," one who can feel the dogs' needs, and understands dogs to the point where they can customize the dogs' education to each dog's particular personality. I've watched trainers who have a ballet precision to their movements. I've marveled at their sense of timing and anticipation of what the dog will do next.

At the same time, many trainers are new, sometimes poorly trained, or in training for the first time. Many are overworked. They get the dog out of its kennel, put it through its paces, and put it back. Eight dogs a day, five days a week, two weeks' vacation a year. They never heard the term "instrumental conditioning," have never read more than part of a book on dog training, and have an anthropomorphic impression of what learning is all about.

Most agencies strive to meet the goal that no dog should graduate with less than sixty hours of training. If the trainer has a string of eight dogs, and works an eight-hour day, it means each dog couldn't possibly get an hour a day, what with necessary breaks for the trainer. But an hour a day, five days a week, for twelve weeks gets you a dog that has met the minimum standards.

At the end of training the dog is matched with its final owner. Who that owner will be is a topic of its own. Often, the owner is inexperienced and/or inept with dogs. He or she may never have had a dog, never trained a dog, never read a book on training a dog. Most don't have ballet-precision movements or understand that consistency of movement and command is necessary.

Again, the dog goes to a new environment, to an owner who did not train it. In fact, the dog might never have been directed by someone sitting in a wheelchair. Believe it or not, dogs and other animals get conditioned during the critical period to signals on a horizontal plane and signals on a vertical plane. Cats raised with vertically slatted spectacles cannot avoid horizontal objects later in life. Lions do not recognize a person in a car. Sheep learn about vertical people during the critical period. Dogs are very sensitive to body language and facial expressions, and easily recognize what they know. But if the situation is unfamiliar, the dog must adjust. I've had my own dogs bark at me as I sit in my car, as if I were a perfect stranger. If I speak to them they look embarrassed, as if to say, "Sorry, boss, I couldn't recognize you through a car window because that is not the way my brain works." The dog has to learn from scratch. The wheelchair-pulling dog has to learn to recognize a command from a sitting person, and how to interpret an unclear signal.

I've watched these dogs work. Like everybody else I was impressed with what a dog could be taught to do. The dog might be the only species in the world that can be trained to do any of these tasks. Try to imagine a guide cat, or a bomb-sniffing pig. Except for the therapy animals, it is hard to imagine any other animal performing any one of these assistance-dog tasks.

However, the fact that dogs can be trained to perform does not mean that they benefit. The tasks these dogs perform are complex, difficult, and stressful. When I watched wheelchair dogs at work in a class, I thought the tasks they were asked to do were so hard that the dogs couldn't really accomplish them. Many of the dogs were terribly overweight. Even a small amount of work drives up the internal heat load. But these were not fifty-pound sled dogs in racing shape, running on a cold day on soft snow. These were seventy- to eighty-pound golden and Labrador retrievers on a sunny, hot California day on black-asphalt pavement. These were not dogs that were carefully fitted with a proper harness that did not restrict the movement or hurt when the dogs pulled. These dogs were wearing poorly designed breastplate harnesses seemingly patterned after those used for horses, with straps lined with imitation fur (more heat) and fitting improperly. The harnesses—all of them—sagged down over the dogs' shoulders in such a way that they pulled against their front legs.

To make matters worse, the dogs were attached to the side of the wheelchair, which forced their bodies out of line. They arched their backs in an awkward position, trying to keep the chair from rotating away from the direction of pull, and at the same time they tried to keep their little toes from being run over.

Sled dogs get tired, even exhausted. But the only hurt they should feel is that almost pleasurable ache that every champion feels from inside the tired muscle. These wheelchair dogs looked like they hurt from the outside in.

Pulling a wheelchair doesn't look hard. Certainly it isn't too hard to

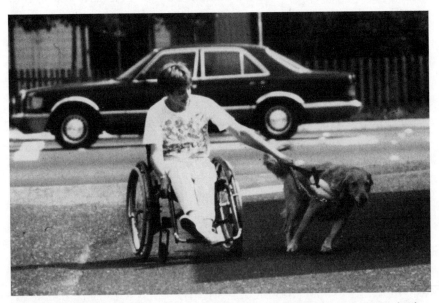

Assistance dog pulling wheelchair. This is a difficult task for this dog and it is doubtful that either the dog or the person can keep it up for very long. Certainly the dog could not go marathon distances day after day, walking on the sides of its feet with its back twisted. Sooner or later the dog will hurt itself. Even the harness doesn't fit correctly, riding over the dog's forearm. It would be like climbing a mountain with a heavy backpack strapped to your lap.

push one (on the flat). But try pulling one from the side up a slight incline. Pulling at an angle makes the task many times more difficult. It not only hurts, but it does damage. It pulls bones out of line, chafes skin, and wears feet unevenly.

Why would a dog want to do this? Sled dogs run to near-exhaustion because it is socially rewarding to run and be with other running dogs. They grow up socialized with other dogs. They watch teams of dogs coming and going. They clamor to go. They don't ever want to be separate from other dogs and left behind, alone. That is the motivation for running and pulling in the team.

Are wheelchair dogs even cognizant of what pulling the wheelchair or pulling open a door is for? What is being accomplished for the dog? The owners usually think that the dog understands that if it opens the door, the person can go through, or if it pushes the light switch, the light will go on. There is no evidence whatsoever that the dog is aware of what is being accomplished. When the dog is pulling a wheelchair, someone keeps saying, "That's a good dog, that's a good dog, that's a good dog, that's a good dog." When the person stops saying "That's a good dog," the dog can quit. What's the reward for performance? That someone stops saying "That's a good dog." Silence is the reward.

Service-dog people keep telling me they need better dogs. They say, "We need smarter, more intelligent dogs." Design us a better dog, has been their request. And there it is again, the concept of robot-dog produced by good genes. I know some people believe that scientists don't credit dogs with feelings or thoughts, and certainly we do try to avoid anthropomorphizing animals. But as a scientist, I don't think a great companion dog can be produced with eugenics. I don't think great dogs are the product of instrumental conditioning. I doubt that a great dog can be fashioned like a pigeon in a Skinner box.

What I learn from watching potential service dogs is that the agencies don't need smarter dogs. They don't need better dogs. They just need to focus more on every aspect of their development. They need to rely less on genetics and more on biological studies of behavioral development, especially during the pups' first year. They need long-term trainers who can get to know the pups from birth all the way to placement, with regular follow-up on the adult dogs as canine assistants. When Charlie Belford, a champion sled dog driver, looked at a new litter of pups, he would use his experience from other dogs, other successes, and envision the various pups two years down the road, harnessed in the team for a race.

"See that gray one?" he would say to me. "That's my future lead dog." Whether or not that turned out to be true, the dog received constant attention, which increased its chances of becoming a good working dog. The situation recalls the golden puppy, Nous, who became a talented working dog and the founding sire of golden retrievers.

The advantage held by sled-dog drivers and hound people is they can see, within the space of two years, how a pup turns out. They use accumulated wisdom to adjust the pup's progress through puppyhood and into training. If they make a mistake, they see the consequences. Everything Belford did with his pups was directed specifically and minutely at fostering them for their adult success on his dog team.

Belford may not have known about critical period as a biological event, but he certainly understood the results of his dogs' early experiences. Starting with dogs that had the structure for running fast and long, he made sure they were never asked to perform beyond their current ability. He kept them in their adult environment right from puppyhood, so that running in the team was familiar, it was internally motivated, and the reward was simply in the doing.

Service-dog people need to understand the difference for dogs between what Belford did and what agency employees do. They need to understand how important it is for livestock-guarding dogs to be raised with livestock, and for walking hounds to grow up in a hunt. They need to apply that understanding to their own industry. Of all the working dogs, service dogs have the most difficult jobs, and the least fun while doing them. Theirs is a stressful occupation with little reward. Many are simply sterilized workers, or slaves, with little to no biological benefit for performing well. Theirs is a dead-end occupation. Their symbiotic relationship with people is thus dulosis—slavery.

I believe the relationship could be, and should be, changed. The opportunity in this field for enhanced dog-human relationships is exciting. The tools and knowledge are widely available, but they need, first of all, to be understood and adopted, and second, to be updated and improved constantly. Both dogs and people will benefit.

PART IV

THE TAIL WAGS THE DOG

In Part IV, I'll try to sort through what's been said in the first three parts, and synthesize for you why the dog's biology is so important for understanding what a dog really is. This book points out why the dog's looks and behaviors are different from what many people think they are. Among the new explanations here are that ancient people didn't domesticate the dog from the wild wolf, and that most dogs are not man's best friend nor are most people dog's best friend. People and dogs live in a symbiotic relationship that for the dog is obligatory. Although popularly considered to exhibit a strong, idealistic mutual relationship, rarely do dogs and humans provide for each other the necessary biological benefits that would rate the name mutualism.

To further emphasize the distinctiveness of dogs, I dallied with Darwin for a bit, and debated with him about his reasoning on the evolution of the dog. Darwin tended to start his arguments in the middle of an evolutionary sequence and move backward and forward from that point. He assumed like many of us that since the people around him purposefully bred dogs now, they must also have done so in the past, and that all dogs are a product of human endeavors. That assumption leads people to believe that the little mongrel dogs running around the streets of the world are escapees from purposeful breeding programs. The fact that many of these dogs running around the streets in Western countries are escapees from our conscious efforts to achieve purebred dogs muddles our understanding of the initial evolution of dogs. But it just may be that these village scavengers are the products of natural selection (not artificial selection). Present-day village dogs live in astronomical numbers around the world. Their very presence is telling us that there is a niche out there that dogs can occupy very successfully. They are giving us clues, by their looks and behavior, about how and when dogs evolved and are continuing to evolve.

Why these new approaches are so necessary becomes obvious in the modern imperative of worldwide management of wild and domestic species. I will need to explore these ideas a bit more. At the same time, I think I can show connections between the true dog and some of the current controversies about it. For example, some scientists believe the dog's scientific name, *Canis familiaris,* should be changed to *Canis lupus familiaris,* to reflect its wolf ancestry. There is also a discussion about just when the dog, as dog, appeared, with one study suggesting the dog evolved ten times earlier than previously has been thought. And there is always the question of how it happened that an ancient and biologically conservative wolflike animal changed into so many sizes and shapes and colors in such a brief instant of evolutionary time. How the plastic dog evolved is not only a question of who did it, but also how it was biologically possible to stretch the dog into so many new shapes so quickly. How can it be that there is less difference in head shape (shape, not size) between wolves and the generic dog than there is between borzois and bulldogs?

For many a dog lover, these issues are confusing. How can Chihuahuas and mastiffs, which probably cannot breed with each other, be the same species; and dogs, coyotes, and wolves, which can breed together, be different species? Since dogs and wolves can and do breed with each other, why shouldn't we think of the dog as a wolf?

Thus, for the final chapters I would like to decontaminate some of the ideas about dogs. It is not that I can answer all the questions about dog evolution, but rather I can at least frame the discussion the way a biologist would, to sort through the questions.

CHAPTER 9

What's in the Name
Canis familiaris?

W̲HEN A BIOLOGIST WRITES a scientific paper for a professional journal, the first time he mentions his study animal, by convention,
he gives the Latin designation of the genus and species. For example, I
start the Introduction of this book with the domestic dog, *Canis familiaris.* The domestic dog has worn this original Latin binomial designation ever since Linnaeus bestowed it in the mid–eighteenth century.

From some (unknown) time in the past, dogs have changed from
wolflike shape, color, and behavior to dog shape, color, and behavior. We
know the ancestor was wolflike because dogs have basic characteristics
that are identical to wolflike animals. The number and shape of the teeth
are "identical" to those of wolves, coyotes, and jackals. Dogs and wolves
have the same number of chromosomes, and the same shape of chromosomes, and the same ordering of genes on those chromosomes.
There is no appreciable difference in the genetic material of dogs,
wolves, coyotes, and jackals.

At the same time, dogs have diverged, changed, transmutated from
their wolflike ancestors. Dogs differ from the other canid species in
measurable ways, just as coyotes differ from wolves in measurable ways.
But dogs win the prize for being the most measurably different and
diverse member of the genus.

I keep saying that dogs evolved from a wolflike ancestor. Why don't I
say that dogs evolved from the wolf? (Indeed, in most of this book I did
say dogs evolved from wolves. Each time I did, I cringed slightly because
it is not quite right.) Present-day wolves and dogs shared an ancestor. For

many biologists, that common ancestor is extinct, by definition. Present-day wolves may look or behave more like the wolflike ancestor, but present-day wolves have evolved just as much away from that ancestor as have the dogs. Perhaps a clarifying illustration is wolves and coyotes. Wolves and coyotes also evolved from the same ancestor. But wolves evolved in Eurasia and coyotes evolved in North America. They evolved at different times from the same ancestor. Wolves later migrated to North America. So which is the ancestor of today's wolf or coyote? Neither—the ancestor is extinct.

The genus *Canis* is a biological phenomenon. Imagine that the same set of genes can produce all those different shapes, sizes, and colors, from the African golden jackal to the arctic gray wolf to the Tibetan Lhasa apso. How the same set of genes code for all those different shapes is the subject of Chapter 11. In this chapter, it is enough to be amazed that there is no appreciable genetic difference between the golden jackal and the little Tibetan lapdog. The same genes produce the long-nosed borzoi and the short-nosed bulldog.

Each of the species and breeds within *Canis* has a sixty-three-day gestation period, and all of them from the largest wolf to the thirty-five-pound border collie have the same size and shape pups. Whether a Saint Bernard or a coyote, pups average around three quarters of a pound. (Of course, there are always exceptions, and the toy breeds do have smaller pups. But the differences in pup size are not as great as the differences in adult size.) Often, differences in pup size are the result of how much energy the female can invest in the litter, rather than genetic differences.

Puppy size is not only similar across the entire genus, but the shape of the puppies is practically identical. (I'll discuss a couple of exceptions in Chapter 11.) The theory is that the "standardized" size of the canid puppy is an adaptation to suckling. Evolution adapts the adult wolves, coyotes, and jackals to different niches, but the puppy niche is identical for each of the species, and selection for that suckling niche keeps their little heads the same highly adaptive shape.

Besides size differences, geography and culture keep these different species from breeding with one another. Coyotes and jackals rarely breed together because they are separated by continents and oceans. Wolves live in the forest and dogs in the city, which reduces their chances of meeting one another in amorous ways. But they do meet at the margins of civilization and wilderness (like the dump), and they can (and do) breed together there.

But even where these species overlap, they don't often mate together. Their social behavior also keeps these species reproductively separate. A wolf pack is very conservative about accepting new members—even new wolf members. Females are most likely to get bred by members of their immediate group. Similarly, five or six livestock dogs guarding a flock of sheep in some faraway pasture protect their females from foreigners, be they wolf, coyote, or even strange dogs. Thus, because each individual is born into a social group, which excludes strangers, their genes tend not to stray. The result is that when all proceeds normally, the groups remain pure. Disruptions in the social grouping, however, in regions where species overlap can lead to outside-the-group matings, as long as other problems (such as size differentials) don't discourage them.

But the lone wolf, jackal, coyote, or dog will form new social groups and reproduce with the nearest lonely somebody else. With all the wolf-killing programs worldwide came an increase in the number of stories of lone wolves teaming up with dogs. Some of these pairings have resulted in offspring. With the breakdown of the normal stability of flock-shepherd-dog while on migration, flock dogs can get lost, and local dogs can intrude upon the disrupted social schemes.

Some conservation biologists feel that the demise of wolves in the lower forty-eight of the United States has led gray wolves to hybridize with coyotes along the Canadian border. And red wolf restoration people worry that reproductive relationships with coyotes may have already precluded ever finding a pure red wolf.

Since they all can interbreed, that brings us back to our initial question: What should the scientific name of these animals be?

Are red wolves *(Canis rufus)*, gray wolves *(C. lupus)*, coyotes *(C. latrans)*, dogs *(C. familiaris)*, and jackals *(C. mesomelus, C. adustus, C. aureus, C. simensis)* all different species? The fact that they are genetically so close creates infinite identification problems for scientists. In the 1960s a wild canid began to appear in New England. Most individuals are smaller than typical wolves, more like coyotes, but they are not typical coyotes, and have some of the characteristics of wolves. They do not sweat through their feet, for example (wolves don't either, but coyotes do). There are dozens of theories about what they are and where they came from. Some think they are coyote-dog crosses, or coyote-wolf crosses.

Maybe we have such trouble identifying these species because they are misclassified in the first place. Perhaps they were never real species, but rather races, varieties, or subspecies of the same species. Darwin thought of subspecies as incipient species, meaning they are on their way to evolving into new species but haven't crossed that imaginary line between subspecies and real species. Or maybe they started to speciate, adapting to different niches sometime in the past, when climate and habitats were different. But now times have changed, and the process has reversed, and they are all blending back together again as they adapt to the present world.

The classic definition of species that we all learned in high school was stated by evolutionist Ernst Mayr, who wrote *the* book on the subject in 1942. In *Systematics and the Origin of Species,* he defined species as "a group of actually or potentially interbreeding populations that are reproductively isolated from other such groups." Wolves, dogs, coyotes, and jackals can all interbreed, or potentially interbreed, and produce viable offspring. Therefore, they are not "good" species (according to Mayr's definition).

How did wolves and coyotes get to be classified as separate species in the first place, if indeed they are capable of interbreeding? It turns out that the concept of species has had an evolution all its own.

In the 1730s Carl von Linné, a Swedish botanist and taxonomist, to whom the world commonly refers by his Latin name, Linnaeus, devised a binomial system of nomenclature and proceeded to give distinct Latin names to "all" the plants and animals. He believed that God had created the different species as part of the original creation. And, he believed that God had given him the task of naming them in Latin. Why Latin? It was the language of Linnaeus's religion. Educated people in Linnaeus's world all studied Latin in high school, regardless of their native language. Latin was a universal language.

For the domestic dog, Linnaeus chose the name *Canis familiaris*. His binomial system designates *Canis* (the genus) as a noun, therefore capitalized in Latin, and *familiaris* (the species) as an adjective modifying the noun, and never capitalized, even if it refers to a proper name. We put the letter L. after the name if Linnaeus gave it its name. The dog is *Canis familiaris L.*

End of story!

Not quite. During Linnaeus's time, when Europeans were exploring the world and sending newly discovered species home to the museums, they needed a system for naming all their new species. Linnaeus developed a system for naming organisms, and developed a key for identifying unknown species. If I found myself in outer Mongolia looking at an animal skull and didn't know its Latin name or even if it had a Latin name, I could consult the key. Linnaeus arranged his list of names such that animals with similar characters had similar names. All those mammals with thirty teeth he called felids, and all those with forty-two teeth he called canids. Since both the wolf and dog have forty-two teeth, he started both their names with *Canis* and listed them together in his book.

It wasn't that he thought dogs and wolves were related, but rather that God had created dogs and wolves with the same number of teeth. But there were not only similarities between dogs and wolves—there were differences. Linnaean species are now called *morphological* species. That means the species differ in some measurable way. I can take a skull, measure the brain volume, tooth size, and head size, look these up in the key, and identify the skull as belonging to a dog. Another skull I might classify as a wolf's, based on differences in those same traits. In this system, dogs, wolves, coyotes, and jackals are different species simply because their skulls, bones, and skins measure differently.

A neat example of this process occurred in the discovery of fossil remains of what could be dogs. Archaeologists Simon Davis and François Valla found a canid tooth in an ancient Natufian village and measured it. They reported that the tooth was an average size for that of a dog, and probably too small to have belonged to a wolf. Therefore, they said, this could be evidence that dogs lived in these villages, twelve thousand years ago. The fact that the tooth was found in the village, an unusual place to find wolves, and was not the average size of a wolf tooth, provides circumstantial evidence for the presence of dogs. They also found buried together the skeletons of a woman and a canid puppy. But because wolf puppies and dog puppies have no measurable differences, one can't tell whether this woman was buried with a choice lunch to take her to the happy hunting grounds or whether she was taking her pet puppy to heaven with her. Also, at that time in history, wolves and other wildlife tended to be smaller than they are now. One has to make the argument that there appears to have been human activity at the same time as the existence of a species of canid, and this canid could have been and might likely have been the dog. The evidence is suggestive, but not proof. Is it suggestive of the first dogs? More on this subject in Chapter 10.

Linnaeus's system of morphological classification (naming) has worked fine ever since he created it. When I went to college, my professors taught me how to use the key, and said I should be able to identify everything with its help.

Still, the story is not complete. Fifty years after Linnaeus's publication, in the same year that Darwin was born (1809), Jean Lamarck published *Zoological Philosophy,* the first book to suggest clearly that species evolve. Species, Lamarck wrote, can change from one shape to another.

Linnaeus (like most scientists of the time) thought that species were created in a permanent shape by God. Culturally, many of us still tend to believe that species are immutable. People argue for species preservation. Dog breeders believe the form of their breed is ancient, and should not be allowed to change. A woman told me once that Anatolian shepherd dogs are seven thousand years old and I had no right to crossbreed them. Breeders have dog shows, at which their dogs are measured to determine which one is closest to the breed standard. The breed standard, they aver, is what the breed is *supposed* to be! Breeders have a key just like that of Linnaeus. For breeders, breeds are morphological breeds.

The idea of morphological breeds or species is older than Plato. We refer to dog breeds and species as if they had a Platonic form. By the right hand of God is the perfect rottweiler, and our earthly job is to make all rottweilers look like the rottweiler that God created. We have an image of what the perfect dog looks like, a Platonic form. (I think one reason we have trouble accepting the village dog as the original dog is because it is very hard to visualize it as the Platonic form of the protodog.) Surely, one would surmise, the very first dog should have some kind of wolflike shape.

Our story of how the dog got its Latin name takes a turn for the worse in 1858. One hundred years after Linnaeus's *Systema naturae,* Alfred Russel Wallace and Charles Robert Darwin presented back-to-back papers to (ironically) the Linnaean Society, suggesting a mechanism by which the Lamarckian transmutation of form (evolution) could occur. Species change, they believed, from one form to another gradually, by a natural selection of the fittest individuals. Because some shapes are better than others at surviving, then those individuals would leave more offspring to the next generation. Each new generation would be slightly more efficient in their economies. And each generation would resemble the survivors of the previous generation. For Darwin, adaptation and speciation are part of the same process.

Interestingly enough, the discovery of the transmutation of species by natural selection did not create much of a problem for Linnaeus's binomial system of naming species. The reason was the key. Darwinian evolutionists came to realize that dogs, wolves, jackals, and coyotes were all descended from some common ancestor, which had forty-two teeth. They all had forty-two teeth not because God made them with forty-two teeth, but because they all had a common ancestor with forty-two teeth. Since Linnaeus had listed all the animals with forty-two teeth together, his key worked for creationists and evolutionists as well. Thus, most names given by Linnaeus have stayed the same since 1758.

In 1982, J. H. Honacki and colleagues suggested that Linnaeus's name for the dog be changed to *Canis lupus familiaris* to reflect its evolutionary descent. Not only did they think the dog is descended from the wolf, but given its similarities to the wolf, they were suggesting that the dog is not a distinct species, but rather a subspecies of the wolf. The Darwinian theory portrays subspecies, races, and varieties as incipient species, or on their way to becoming real species. Since all these canids are interfertile, they are probably not very far along the way into branching off into "real" species.

What Honacki and his colleagues were saying is that the dog has not gone far enough down the path to be counted as a real species. They are also saying that the dog is descended from the wolf. In one fell swoop, the wolf-changed-into-the-dog believers win the long battle of who was the progenitor of the dog.

Personally, I have no problem with the idea that dogs and wolves shared a common ancestor. To try to reflect this according to established principles of nomenclature, however, has led to an extraordinary convolution and distortion of the interpretation of canine taxonomy. A few years ago I was working on the New Guinea singing dog with zoologist Lehr Brisbin. These are a "wild," "feral," or perhaps "free-ranging" dog from the island of New Guinea. They live in the forest and sneak into the villages at night and scavenge on human waste. We were writing a paper on our findings, and we had to designate our animal with the accepted Latin binomial, so other scientists would know just which animal species we were working with.

Brisbin called up one day and unfolded the following mess. In 1957, Australian mammalogist Ellis Troughton had reported the "discovery" of the New Guinea singing dog. He thought it was a separate species and designated it with the Latin binomial *Canis hallstromi*. This is a perfectly good name, but many felt that the name did not represent the ancestry accurately, and thus a tortuous renaming process began and is perhaps still in progress. Some experts felt it was just a dog and referred to it as *Canis familiaris hallstromi*, a subspecies of dog. But dingo expert Laurie Corbett thought it to be a variety of the dingo (which used to be *Canis dingo*, but has been reclassified *Canis familiaris dingo*). This made the New Guinea singing dog *Canis familiaris dingo hallstromi*. But then J. H. Honacki and his colleagues in their 1982 taxonomic and geographic reference suggested the domestic dog should not be *Canis familiaris* but rather *Canis lupus familiaris*, a subspecies of the gray wolf. Technically, then, that would make the New Guinea singing dog *Canis lupus familiaris dingo hallstromi* or a subspecies of dog, and a sub-subsubspecies of wolf.

So why does it make a difference? Who cares? It matters because the species designation has legal implications. In our coauthored paper on the New Guinea singing dogs, Brisbin argues that the dogs are rare and endangered and should be protected. If we call them *Canis hallstromi*, it would be easy to make the case for protection, since they would represent a specific form that could be measured, described, and protected. But if they are a subspecies of an Australian crop pest (the dingo), which is a subspecies of the most common canid in the world (the dog), which is currently a subspecies of a regionally threatened species (the wolf), then we are in big trouble trying to explain to our congressman why we think New Guinea singing dogs are a rare and endangered "species." Brisbin would like to see the designation changed to *Canis lupus hallstromi* for just those reasons. That is fair enough, but I think *Canis hallstromi* would be better. *C. lupus hallstromi* just takes the name back to a pseudogenealogical status, which is not the point. What's important is which population of animal we are trying to describe. It doesn't matter what their ancestry is when you are trying to protect them. Why should we try to combine political events (protection of a species) with favorite hypotheses about the origins of a species?

Ecologists have a good way of thinking about species. They think of species as a population of animals that is adapted to a niche: one species, one niche. No two species can occupy the same niche, is an ecological rule. Dogs, wolves, coyotes, and jackals are each adapted to different geographies, different environments, different habitats, and, ultimately, different niches. Each of these species earns its living in a different way. Wolves kill big prey, and it is assumed that their large size is an adaptation for pursuing and capturing big animals. Coyotes have a size and shape economical for killing rabbits. Jackals are jacks of all trades, quite often scavenging the remains of other predators' kills. And as I pointed out in Chapter 1, dogs have little heads, little brains, and little teeth, which are economical adaptations for a village scavenger. Dogs have an obligatory symbiotic relationship with humans; nobody else in the genus does. Therefore dogs are a species.

Canid species may occasionally breed together at the margins of their niches. But the ecological theory says these will be infrequent occurrences because the offspring or hybrid of two different ecotypes, i.e., individuals adapted to different environments, would not be adapted to either of the parental niches. Therefore the hybrid between a wolf and a coyote could not kill either big prey or small prey as well as do the "pure" forms. The hybrid between wolves and coyotes would have less chance of surviving than its pure cousins because it is not adapted to any niche. The hybrid is not really viable.

Thus, ecologists would classify dogs, wolves, coyotes, and jackals as different species because they are adapted to different niches. This system meets the spirit of Mayr's definition of sexual isolation. The dog-wolf hybrid could not compete reproductively in the wild, or in a symbiotic relationship with people.

The English poet John Donne understood the problems a hybrid might have, describing the fate of a pup whose mother, a dog, had been bred by a wolf.

> *Being of two kinds thus made,*
> *He, as his dam, from sheep drove wolves away,*
> *And, as his sire, he made them his own prey.*
> *Five years he liv'd, and cozen'd with his trade;*

Then, hopeless that his faults were hid, betray'd
Himself by flight, and by all followed,
From dogs, a wolf, from wolves, a dog, he fled,
And, like a spy to both sides false, he perished.
"Metempsychosis," verse 45.

As a behavioral ecologist I regard the dog, *Canis familiaris,* as a separate species, a product of a distinct evolutionary event. The dog is beautifully adapted morphologically and behaviorally to feed and reproduce efficiently in the company of people. Wolves are very awkward in a people environment. I see wolves and dogs as adapted to different niches, and so, in my brain, they qualify as different species.

If indeed the science world does insist on renaming the dog *Canis lupus familiaris,* there is one thing we must all remember: Just because dogs are renamed as a subspecies of wolves does not make them wolves. To say that dogs are descended from wolves does not make them wolves.

To say we are descended from apes does not make us apes. To say we have 99 percent of the same genetic makeup as chimps does not mean we should raise our kids as if they were chimps. To say dogs have 100 percent the same genes as wolves does not mean we can treat them as if they were wolves.

Dogs are not wolves, no matter what you call them.

CHAPTER 10

The Age of the Dog

I N PART I, I TREATED the dog as an evolving species, perhaps twelve thousand to fifteen thousand years old. I believe that biological and paleontological evidence supports the designation of the dog as a product of natural selection—that common anatomical dog features are special adaptations for scavenging human waste products in the new niche that was initially created by a new human social organization, starting in Mesolithic villages. Therefore, the dog cannot be older than these relatively permanent settlements.

Actually, I really think dogs as an adaptive form don't come until much later, in the Neolithic age. To me, dogs, like the other domestic animals, are the products of serious agriculture. The evidence of dogs twelve thousand years ago is scanty, and not much better for the next three or four thousand years. But by the time agricultural communities become established, we see that livestock-guarding dogs and walking hounds were firmly rooted. But because I also consider these agricultural dogs to be evolutionary by-products of canid village scavengers, I have to remain open-minded about pre-Neolithic origins.

That is my working hypothesis. There are other hypotheses which other investigators believe are also plausible, claiming that different data support their ideas. Fair enough. Our scientific trade is to accumulate data that support (or deny) a hypothesis. If someone can show that the dog is older than permanent settlements, or that they were a result of pet keeping, or the result of generations of tamed and trained wolves, then I would be required to adjust, adapt, or abandon my hypothesis.

But in the meantime there is an important biological question much more serious than just when the first dog evolved. No matter when the first dog evolved, it appears to have happened very fast. One reason the time involved in creating dogs is so important is because of implications for Darwin's theory of evolution. Even if dogs did diverge from wolves 135,000 years ago, as a recent article claims, that is still startlingly recent in terms of Darwin's theory of artificial and natural selection. What we need to do here is investigate the possible time frames for domestic-dog existence and then devise a model of how the dog's evolution could have occurred in that time.

There is no such thing as the first dog, nor would such an animal be identifiable. Even finding the first permanent settlement might not be possible. I can imagine that the first permanent settlements were grass-house communities, existing by the sea at the edge of the continental shelf, during the glacial periods. These villages could have existed for thousands and thousands of years before the melting glaciers and the rising sea level destroyed all evidence of them.

Normally we think of permanent settlements as houses of stone. "Permanent" is an archaeological term which implies that the buildings were so solid, evidence of their existence lasts forever. As far as the original wolf-to-dog scavengers are concerned, however, all they needed was a settlement of people existing by a fabulous fishery or a shellfish lagoon, or some other kind of natural animal trap, for a long period. The scavenger model depends on some kind of continual dumping of human waste. I should change the term from "permanent" to "continual" settlement, and define it as a place where people discarded their waste in one spot for many generations.

So far, in paleontological investigations, the first houses of stone appear perhaps twelve thousand years ago in the Fertile Crescent of the Middle East. The oldest known villages belonged to people called Natufians, part of a culture that extended throughout what is now northern Israel and the surrounds. These villages are described as Mesolithic, meaning that the people were hunters and gatherers, not agriculturalists. Hunters and gatherers are people who wander around looking for food, but they can also hunt and gather from the same spot. The Pembans have hunted and gathered on the edge of a coral reef for generations, bringing their produce back to continual settlements of not very permanent houses.

There is some evidence that people dwelt in continual settlements twenty thousand to thirty-five thousand years ago. These people lived or spent a lot of time in caves in France and Spain. We know this because they covered the walls with beautiful paintings. The mouth of one of these caves is under the sea, which is good evidence that humans occupied a portion of the continental shelf before it was eventually flooded by a rising sea level as the glaciers melted.

The Chauvet and the Lascaux cave paintings are significant for dog investigators for several reasons. Most notably, no domesticated animals are pictured, including dogs. It seems to me if they had had dogs and used them for hunting or even if they were esteemed pets, these gifted artists would have included them on the walls. On the other hand, if there were dogs, but the people thought of them as scavenging pests in their dump, they might not have drawn pictures of them.

The caves are evidence that people then spent long periods living continuously in one place. Could wolflike creatures have been evolving into dogs in this period? I don't see any biological reason why not. If Belyaev's observations on the transmutation to tame foxes apply to dogs, then all I need is a place with high selection pressure for half a century and, presto! I could have the protodog.

What are the chances humans were living together in continuous settlements forty thousand years ago? When we go back forty thousand years, we are running into that sticky question of whether we think *Homo neanderthalensis* was human and living in permanent settlements. Conjecture follows conjecture; we make educated estimates. But we will just have to wait for the discovery of more information.

A paper published recently in *Science* magazine suggested that dogs diverged from wolves about 135,000 years ago. If humans had any role in the domestication of dogs, then 135,000 B.P. is essentially the *Homo neanderthalensis* era. The concept begins to push the limits of what is known about humanness. Could people talk? Did they have the cognitive abilities of abstract thought? Could they draw abstractions on cave walls? Some experts think the answer to every one of these questions is no.

But could some proto-Neanderthals have hunkered down on a sea coast and created wastes, attracting scavenging wolves and forming a symbiotic relationship with them over a fifty- or hundred- or thousand-year period? Why not? Make up any story you want, because there is no solid evidence for such an early appearance of the dog. The paper in *Science* seemed to present a breakthrough in determining the age of the dog, and indeed the media treated it as such. But since that paper was published, the reliability of mitochondrial DNA as an evolutionary clock, on which the 135,000-years-ago divergence was partly based, has been seriously questioned. Not only do I (and others) perceive significant problems with the methodology and therefore also the results, but I think the media have misinterpreted the results as published.

Fossil evidence of dogs even twelve thousand years ago is slight and circumstantial. Essentially, a very few findings that seem to indicate a close association between people and canids are taken as evidence of domestication. I think that is fair enough for the moment, but what evidence there is should not be confused with conclusive evidence. If there were dogs in the Natufian villages, there certainly weren't very many of them.

By four thousand years ago there were dogs in abundance, but little evidence of identifiable breeds. By Roman times, two thousand-plus years ago, writers are describing both sheepdogs and hunting dogs, and what sound like village curs are variously described in the Bible and other works written less than a thousand years before Roman times.

But, currently, there is no archaeological evidence that dogs existed between 135,000 and 12,000 years ago. At 12,000 years B.P. there are two tiny little pieces of what may be evidence (and may not): the skeleton of a human found buried with a canid puppy (which could have been a wolf puppy) and the smallish adult canid tooth found in a nearby house site. As I discussed in Chapter 9, the tooth was in the range of wolf-sized teeth, but at the small end of that range. The puppy being buried with the human was interpreted to denote a cultural association between the two, or was, perhaps, a sign of affection. And what about the tooth? What was a smallish wolf doing in someone's house? My house is full of wolf teeth but there has never been a live one in here. Could some little kid have lived in the house and had a wolf tooth as a lucky charm?

Is there any other evidence of dogs from that 12,000 B.P. period? Yes, but always it is confounded by the difficulty of telling which species a puppy skull belonged to, wolf or dog. Wolf puppies are virtually identical to dogs of like sizes up to five months old.

But so far there is zero evidence for dogs before 12,000 B.P., including the recent mtDNA study suggesting a 135,000 B.P. origin. This recent paper suggests that domestic dogs could be 135,000 years old. But it could also say that the wolf stock that dogs descended from has been separated from other wolves for 135,000 years. If the latter case is true, the predog wolf stock could have existed, looking and behaving like normal wolves, for 123,000 years, not evolving into actual dogs until people provided the domiciles for them to domesticate themselves—twelve thousand years ago. Mitochondrial DNA cannot reveal which of these two possible scenarios is correct. Since the second suggestion is supported by archaeological evidence, one would have to say that the most parsimonious interpretation, given all the evidence, is that the dog is approximately twelve thousand years old.

How good are the mtDNA data that suggest the ancestral wolf was isolated for 135,000 years? Since that paper was published, many experts have abandoned the mitochondrial clock as a valid estimator of time. They have searched for other clocks, such as the Y chromosome traveling between father and son. Y-chromosome clocks give different results than mtDNA clocks. Either the molecular clocks are notably inaccurate, or perhaps we don't know yet how to read them. My opinion is that right now, within the canids, mtDNA is reasonably poor at determining genealogical relationships, and even worse as a clock.

When I studied the dogs on Pemba, an island with limited immigration and emigration, I was impressed with the effects of disease on the population of dogs. Their numbers would get very high and then crash. Diseases that kill are a phenomenon for every species, including wolves and dogs. When the population crashes, mtDNA information, the basis of the clock, is lost. After a crash, the Pemba dogs might have only one or two haplotypes left. (Haplotypes are the mutant form of the theoretically original mtDNA.) If, for example, we suppose these are the "original" dogs and that they evolved on Pemba twelve thousand years ago, as the mtDNA clock ran it would generate more and more mutant haplotypes. After twelve thousand years, one would expect to find many mutations. The original Eve's mtDNA would mutate at some rate in each succeeding generation. We would expect that studies today would find the greatest number of mutations present in the population. But disease and other forms of mortality filter them out, and after a population crash or two, only a very few mutations would be present. If all the dogs on Pemba died except one pregnant female, then she would become the Eve mother of all future dogs. With little to no variation in haplotypes, then the measurement of time or origin with mtDNA is highly inaccurate.

What present studies do is statistically determine the mutation rate given the variation in the surviving haplotypes. The answer is a *statistical* probability.

When I first showed the *Science* article to my colleague, geneticist Lynn

Miller, his first response was "Where are all the wolf haplotypes? Why are there so few?" Where and why, indeed. Is it because we have been exterminating wolves worldwide for so long? Is it because they are afflicted with mange, rabies, distemper, and persecution, and the expected variation in all those haplotypes has been eliminated from the population? The same paucity of haplotypes exists in European humans, who don't show the kind of variation one might expect if the mtDNA clock were running at some constant rate. Was it the wars, famines, or plagues that reduced the variation? Or—perhaps the mtDNA clock is not constant for every population. Or perhaps mtDNA is not a clock at all.

I talked Lynn into teaching me how to read the *Science* paper closely. He shared his criticisms with me, and led me through the paper, sentence by sentence. It was a real lesson in critical analysis. Why were the samples of dogs taken from established breeds, in order to compare with wolves from diverse geographical regions of the world? This method treats breeds of dogs as if they were populations that would represent their country of origin even though most of the dog samples came from the United States and Europe. It seems to me that in a study of the wolves of Romania and western Russia, for example, the dogs used for comparison should be free-ranging dogs collected in Romania and western Russia. Of course, as reported in *Science,* it wouldn't really have mattered, because those wolves had dog mtDNA. Those wolves had the same mtDNA as my Italian Maremmano-Abruzzeses, as the Mexican hairless, Irish water spaniels, and nine other breeds. I single out these three breeds because the Maremmano-Abruzzese is a dog of the transhumance, the hairless is continents away from Russia, and the Irish spaniel is also a founding breed of the golden retriever, whose relatives show up with a number of other dog haplotypes around the world. "Established" breeds are hardly the way to test the genetic relationship between dogs and wolves.

Why didn't the authors use, as is customary, the Latin binomials to designate genus and species of wolves and dogs? They did so for coyotes and jackals. Was it an oversight or a deft deletion, because of that confusing issue about whether dogs are a subspecies of wolves? After all, technically, if they are a subspecies of wolves (remember *Canis lupus familiaris*), divergence is still not as complete as in a speciation. By definition, subspecies are not sexually isolated populations. A subspecies is a nonrandom distribution of alleles geographically based. And breeds of dogs are not species, that is, geographically isolated from the greater dog population, nor are they subspecies in the technical sense, since the founding animals of a breed club are not a random sample drawn from the original geographically based parent stock (the wild stock, so to speak).

The fact that some dogs and wolves have identical haplotypes is good evidence that they are not different species in the classical sense of the term. However, the authors of the *Science* article assumed divergence. They assumed that Linnaeus was right when he classified dogs, wolves, coyotes, and jackals as different species, even though they all can interbreed. Then, when they found that these species carry common haplotypes, they concluded that was because of recent hybridization. If the common haplotypes are the result of hybridization, then the data suggest some of the wolves got their haplotypes from dogs. The Russian wolves are obviously carrying dog haplotypes. That means some wolf packs had dogs for mothers. Which of the other wolf haplotypes were obtained from dogs? And which of the dog haplotypes were obtained from wolves? There is no way to tell. If there is recent hybridization, it implies that it is entirely probable that hybridization has taken place from the very beginning of dogs, whenever that was. It also means you can't distinguish dog haplotypes from wolf haplotypes.

The same problem of species sharing haplotypes appears in an earlier paper from the same mtDNA laboratory, which reported that gray wolves hybridized with coyotes in the northeastern United States, and in another paper, that gray wolves and coyotes were breeding with red wolves.

But there is an alternative explanation to the hybridization hypotheses, which is that current mtDNA techniques cannot tell us the difference among dogs, red wolves, gray wolves, and coyotes.

The question might be framed another way. If I sent an unlabeled sample to the laboratory, could the scientists tell me what species it belonged to? The answer is—probably not. A big problem with the technique, as I see it, is that the sample tissue has to have the species name on it in order to be identified. But its identification is the question, not the answer! How does one know the sample was from a wolf? Well—whoever collected it said it was a wolf. But how could anyone know with certainty it was a wolf? The wolf expert sends an apparent wolf specimen to the mtDNA laboratory, which in turn determines that the mtDNA of the animal is that of a coyote (that is, a hybrid). What does that mean? Was the specimen mislabeled, or was it a really a big coyote, or is it a hybrid, or is it that no one can tell them apart? Isn't it odd that every mtDNA study comes up with the hybridization dilemma? Red wolves are a cross between gray wolves and coyotes, yet in Canada gray wolves are carriers of coyote mtDNA, and 20 percent of the dogs carry wolf mtDNA, and in Africa, the Simien jackal, which might be a wolf, is carrying dog mtDNA. Do these findings really help solve the dog genealogical question?

Lynn Miller began to follow the genealogical argument as it evolved in the literature. More papers and data were published. Then Lynn did something interesting. In his own words: "I did the following search on the NIH sequence comparison machine called BLAST (Basic Local Alignment Search Test). First, I used the Vilà et al. (1997) coyote (about 700-base) sequence and downloaded from the BLAST database the 100 closest matching sequences. Other than two coyote sequences, the next twenty or so sequences are from 'dogs,' mostly from the Vilà et al. database. That would give credibility to their figure 2B, which shows coyotes to be ancestors of dogs. The 100th sequence was a jackal sequence from Randi et al. (2000). The problem with this sequence is, it is only 595 bases long. So the poor match is perhaps due to the shortness of the sequence, at least. I ran the jackal sequence to find its 100 closest relatives. The first closest sequences come from D6a, D6b, and W6, the famous introgressing dog haplotype in Russian and Romanian wolves (that is, the dog/wolf hybrids from the Vilà et al. paper.) These sequences have much more sequence difference between them and the jackal than between them and the rest of the pack, but in both lists coyotes, dogs, and wolves are interspersed together."

I understand these arguments may be difficult to follow. However, I also understand that when the popular press makes statements that the Vilà et al. paper "shows clearly" that dogs are descended from wolves, they are not looking at the data very carefully. Lynn goes on, "Very few of the statistical outcomes in the Vilà et al. paper approach the 95 percent confidence level, which means that these branches ('clades') in figure 2A could occur by chance alone. [Author's note: figure 2A is the one that is supposed to show clear evidence of wolf ancestry.] Only three small branches of figure 2B reached the 95 percent confidence level—and dog clade 1 is not one of them. [This is important, first, because it is dog clade 1 from which both the (species?) divergence and the age estimate are based. Second, figure 2B shows (clearly?) that some dogs are more closely related to coyotes than wolves. And third, figure 2B shows that dogs are older than wolves.] These data led to the comment of another scientist that the standard error here is plus or minus 300 percent. This means, using the same data, that the dog could also be 12,000 years old."

Lynn argues that in order for Vilà et al. to accept the ancient origin of the dog with these data, they would also have to agree that the dog was older than the wolf, as their figure 2B indicates. What the paper really shows (clearly) is that the expected variations in wolf haplotypes, which would give the results some meaning, are missing. From what we know currently about the workings of mtDNA, if wolves have been around for millions of years, we would expect to find a large number of haplotypes. Robert Wayne gave a paper in Spain a few years ago where he reported on mtDNA of a wolf frozen in the ice at the end of the last glacial period (about the time the dog was evolving), and guess what? It showed no relationship to present-day North American wolves. Why? Simply because the mutant haplotypes are lost, constantly, probably for the same reasons I hypothesized for the Pemba dogs, whose population is decimated periodically.

Not only are there problems with the genealogical sampling in the *Science* paper, but the method for determining the ancient dog date isn't all that clean, either. The 135,000-year figure was derived by dividing the gene divergence of wolves and dogs (.01, a figure based on the *not* significant data in figure 2A) by the gene divergence of wolves and coyotes (.075 ± .002), and multiplying the result by the divergence time of wolves and coyotes as established by the fossil record of when that event took place (1,000,000 years ago).

(Don't go away, it is a very simple equation.) The authors solved the equation (.01 ÷ .075 × 1,000,000 years = 133,333 million years old) and concluded ". . . this implies that dogs could have originated as much as 135,000 years ago." Well, yes, if I accept all their assumptions, allow insignificant data in the equation, and ignore the minimum side of the 300 percent standard error, then dogs *could* be that old.

One of the assumptions in the paper rests on the .01 and .075 divergence distance, that is, on the fossil record estimate of when coyotes and wolves split. But how good is that fossil record figure for the divergence? The authors apparently used a 1968 book by Björn Kurtén from which they took the one-million-year figure. But that book really says that wolves and coyotes diverged between one million and two million years ago. That estimate was revised by Kurtén and E. Anderson twelve years later in *Pleistocene Mammals of North America* (1980), to as much as 3.3 million years ago. (And it wasn't a clean divergence, since coyotes evolved in North America and wolves in Eurasia and later migrated to North America.) If Vilà and his colleagues, who wrote the *Science* article, had used the more recent estimate, then their resulting number would have put the dog at between 135,000 and 450,000 years old. One critic tried the same method for the domestic cow, and came up with a figure of 200,000 years old. That would mean the cow is older than dogs and as old as or older than Neanderthal man.

Let's alter the question in the *Science* paper, just for fun. What if its purpose had been to find out when wolves and coyotes—not dogs—diverged? Using the fossil record for divergence of dogs at 12,000 years ago, the equation would read $.075 \div 01 \times 12,000$, or 90,000 years ago.

We could now write a paper that says, "new research with mtDNA suggests that the divergence of wolves and coyotes is much more recent than is shown by the fossil evidence." In other words, the model constructed by Vilà et al. may be a good one, but the answers it gives depend on the starting numbers.

What does it all mean? Mainly, it all means that I don't think the present studies shed any light on the age of the dog or even its ancestry. The species of this genus can't be defined by the Linnaean taxonomic system, which assumes an original creation, an ancestral split, resulting in genetic isolation. When, in these mtDNA studies, two species share a haplotype, it has to be assumed that the sharing is a result of a rare hybridization. This is necessary simply because it has already been assumed that divergence means sexual isolation of the mutating haplotypes.

When I refer to jackals, wolves, coyotes, and dogs as species (Chapter 9), I am not a taxonomist, but rather I have my behavioral ecologist's hat on. I define each of these species by the niche they occupy. I don't think of them as Linnaean species but rather as a single population of animals that has a neat trick for adapting to changing environments. Environmental change (stress) results in diminishing population size and extinction if the organism cannot adapt. Adaptation in the evolutionary sense means evolving a new form. But with small populations, where does the genetic variation come from to select for these newly adapted forms? I think that when jackals, wolves, coyotes, and dogs find themselves faced with failure of their niche to provide enough reproductive opportunities with conspecifics, they just interbreed with their relatives at the margins, and reemerge as rapid adaptations to the new niches.

A few years ago a species of *Canis* was discovered in New Hampshire. Scientists struggled with giving it a Linnaean nomenclature. Forty years later, I'm not sure we know who its ancestors were—coyote, wolf, dog-coyote hybrid, or wolf-coyote hybrid. We have agreed to call it the New England coyote. At least this means it doesn't fall under the Endangered Species Act.

If this kind of ecotyping (hybridization and adaptation) has been going on for millions of years, think of the confusion the paleontologists would have trying to figure out the fossil record. And, if the fossil record is problematic, then one cannot compute an mtDNA date with any accuracy.

I believe that mtDNA studies are an exciting area of evolutionary research. Sooner or later they might reveal a breakthrough on dog genealogies and ages, even if so far they have just added to the confusion.

But at the present time, I think this information about dogs, wolves, and coyotes, as reported and interpreted, has done some real harm. I think wolf restoration programs have suffered, as scientists struggle to devise precise way of defining species and species diversity. The critics of one of the restoration projects simply say, "This is not a real wolf. This is a hybrid with the despicable coyote." How do they know? They learned it from mtDNA studies, which cannot discriminate with accuracy between even wolves and coyotes!

To me, the behavioral ecologist, it is ludicrous to classify the dog and the wolf as the same species. It doesn't matter how closely they are related, molecularly. They are different species, just by inspection of their behavior, anatomical traits, and habitat. The red wolf biologists just shake their heads in frustration when they hear their animal is a hybrid between wolves and coyotes. They observe red wolves, and they see animals that are behaviorally and anatomically different from gray wolves and coyotes.

Similarly, I shake my head in frustration with popular interpretations of the mtDNA studies. About thirty years ago a number of people began to endorse the concept that the wolf was the sole ancestor of the dog. Part of their reasoning was that the dog and wolf behave alike. Dogs are social with people in the same way wolves are social with each other, they asserted. Therefore, they argue, wolves and humans could have made those earliest bonds with each other because of their mutual goals (hunting) and their mutual social solutions to food-catching (packing). The resulting relationship would lead to an evolving symbiont, that is, the dog. It's the Pinocchio Hypothesis again.

The result of this faulty reasoning is a myth that has gained momentum. Dog training books have perpetuated it over and over, with the "wolves form packs" cliché. Then the scientists adopted the myth, and people working with wolves have reiterated it—"Most scientists now agree that the wolf is the ancestor of the dog"—as if it were a popularity contest. Then, to reflect this new agreement, taxonomists nail the conclusion by changing the dog's name to reflect its newly decided status as a subspecies of the wolf.

For me, the dog does not behave socially like wolves, nor are wolves socially anything like people. Wolves are not more highly social than coyotes or jackals, and I do not think there is a ghost of a chance that wolves were domesticated by people to be hunting companions. That kind of homology from one species to another is sloppy, illogical reasoning, based on no scientific analysis and very poor theory. The fact that dogs appear to be more closely related to coyotes in the Vilà et al. paper is not evidence of that relationship—it is just evidence that we should be more careful in our thinking about these matters.

What is most important to me is to investigate how the dog evolved so fast. Darwin and those who have championed his ideas have never been able to show that artificial or natural selection results in speciation in the sexually isolated sense. Indeed, the wolves, coyotes, jackals, and dogs seem to illustrate the differences between species formation and species adaptation by natural selection. A coyote—larger than western coyotes—appears in New England one afternoon in 1963. It is now ubiquitous throughout the region. Anatomically like wolves and unlike coyotes, these newcomers do not sweat through their footpads, and behaviorally they are adapted to New England. Is that gradualism? All the breeds of dogs, represented by all the shapes and sizes thereof, were not created slowly over long periods of time, in the Darwinian sense. They were created, mostly, in the last hundred years. Whatever happened to the dog in the previous thousands of years is minuscule when compared with the divergence of form that has taken place recently. Looking at the divergence of new forms among dogs just might give us an insight into how new forms arise among their wild cousins.

CHAPTER 11

Why Dogs Look
the Way They Do

IF YOU ASSUME, as does J. H. Honacki, that dogs and wolves are the same species, just subspecies of each other, then why is one population of these wolves (dogs) so richly mutable—almost plastic—while the wild type (wolves) is so dull and conservative? Why haven't wolves achieved similar levels of physical and behavioral perfection? Why can't wolves run as fast as some breeds of dogs, or why have dogs better senses of sight and smell than wolves? Why can't you teach a wolf to herd sheep and win a sheepdog trial? Why is it so hard to modify wolf behavior, while dogs seem constantly adaptable?

The domestic dog, as I mentioned in Chapter 10, might well be an ideal "species" for scientists to study the genetics of transmutation of form, behavior, and even speciation. The fact that dogs are so malleable and can become so behaviorally specialized and so physically well-adapted to microenvironments in just a few generations suggests that we ought to study those processes of transformation.

Although some may think that dogs are really wolves with funny colors and floppy ears, they are not. Although some think breeds such as German shepherds or huskies look like wolves, and that some breeds are more closely related to wolves, there is no evidence to support this. "Wolf-dog" is a misnomer—except in the case of a true hybrid, of course. To me, German shepherds or Siberian huskies don't look remotely like wolves; neither do they behave remotely like wolves.

The few genetic studies that have been done do not show any genealogical relationship connecting specific breeds of dogs with wolves. In one study we did with mtDNA, the consensus tree resulting from our statistical analysis showed French poodles to be "closer" to wolves than huskies or livestock-guarding dogs. That is ridiculous. This kind of finding says nothing about French poodles. What it says to me is the mtDNA methodology we used needs refinement in order to be valid.

To infer that one breed is "closer" to wolves, one is assuming either that dogs evolved from wolves many times, and that some breeds have evolved more recently, or one is assuming that one breed has genetically changed less than the others, since the beginning. Neither case can be supported by any of the available information. Perhaps one assumes that German shepherds or Siberian huskies act more like wolves. Nothing could be further from the truth.

When I think of the wild animals existing for millions of years, and all they do is change size a little bit, I have to be impressed with the dog's astonishing manifestations. Six million years of evolution, and wolves, coyotes, and jackals maintain the same relative proportions of skull, teeth, and brain. Six million years of natural selection and survival of the fittest, and they can't bite any harder, run any faster, or see any better. How could that possibly be?

At the end of the Ice Age (twelve thousand years ago), the wild canids, like many of the other mammal species, got smaller. (That's why the small canid tooth reported by Davis and Valla in the Natufian village is questionable as a dog tooth. There were "real" wolves running around Israel then—and now—that weighed less than forty-five pounds.) Perhaps it would be more accurate to say that the big members of the family, such as *Canis dirus,* the dire wolf, became extinct, while the smaller ones flourished. Perhaps the gray wolf is just another big Ice Age relic species going extinct. The theory holds that as the world warms up, the niche that the bigger species occupy disappears. The small species are doing very well.

The coyote and jackal have been separated continentally for perhaps six million years, and yet the side-striped jackal and the coyote have amazingly identical head shapes. Six million years of natural selection and the most innovative changes are represented by the red wolf, the gray wolf, the golden jackal, the black-backed jackal, and that really snappy dresser, the side-striped jackal. If someone were to dig up a side-striped jackal in New Mexico, it would probably be labeled "coyote." A wild canid species showed up in New England in the early 1960s, and experts still argue whether it is a coyote or a wolf or a hybrid of some kind. No good defining characteristics exist except perhaps size, and that often overlaps between the species. Every single character overlaps with the other, "species" characteristic. Even when we examine blood types, protein structures, and mtDNA, they share characters. Geography and size are still the best indicator of species.

Evolutionary biologists expect transformation of form to be slow. Six million years are not a long time in the speciation world. The prevalent theory is that transmutation of species takes place gradually by natural selection. Darwin thought that geological and climate changes were very slow. The environment or climate gradually shifts over time, and the organisms gradually change shape as an adaptive response.

The rapidity with which the dog has changed form, and the seemingly endless varieties of its form, challenge the theory of Darwinian evolution, that adaptation must be a slow process. Whatever the time frame of dog origins (be it 12,000 years or 135,000) doesn't matter, for either pace defies our evolutionary imaginations. Even among the other domestic animals, the number of different shapes the dog assumes is phenomenal. Dogs are a phenomenon. Breeds of cats, cows, or pigs vary in size, color, and coat length, but they don't show the extreme shape differences of the dogs.

There are hundreds of breeds of dogs. Each breed has a definable conformation, a uniformity of size, shape, and behavior—a breed standard, if you will. If we didn't know better we might be tempted to give each breed its own (morphological) species designation. There are larger and more measurable differences between a pit bull terrier and a borzoi than within all the rest of the genus combined.

Within the hundreds of breeds of dogs, breed differences suggest a distinctive genetic program for each. Yet unbelievably, there is no appreciable genetic difference between them. One would think that there was some sequence of alleles that would act as a genetic marker that would identify a breed. Yet to date none has been found. Incredible as it may seem, there is no way one can genetically determine that "this is an American pit bull terrier." I doubt there ever will be a genetic test of breed, which will become clear as I explore the making of breeds.

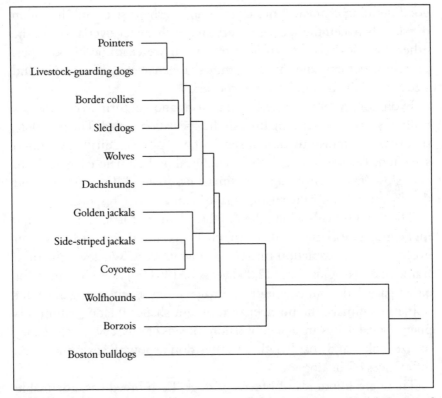

Differences in head shapes. This tree represents the similarities of head shapes of different members of the genus *Canis.* One should not confuse the similarity of head shape with genetic relationship. Borzois and bulldogs are as closely related as any of the animals on the chart but have the most differently shaped heads. Coyotes and jackals have almost identically shaped heads but are probably the most distantly related. It has been argued that since dogs have head shapes much like wolves they must be closely related to wolves. But dog head shapes show more differences between the dogs that they are related to than between dogs and wild types. Head shape is not a good character to try to determine genetic relatedness. (Adapted from an unpublished thesis at Hampshire College by Richard Schneider)

HOW TO CHANGE SIZE

It has become a cliché to start a dog book by writing that dogs are morphologically the most diverse of any known species. The common illustration is the size difference between a two-pound adult Chihuahua and the 130-pound breeds such as the Saint Bernard. I once had an Anatolian shepherd that weighed 178 pounds. The size variation between adult dogs is truly extraordinary.

But perhaps not all that extraordinary. Size is easily adjusted during evolutionary processes. A 130-pound wolf starts as a microscopic egg. At some point in the life cycle of a single wolf, it was identical in size to every breed of dog (except perhaps the very biggest). Newborn wolf puppies are smaller than adult Chihuahuas. In other words, it is relatively easy to play size games with the different canids just by speeding up or slowing down growth. In fact, coyotes and jackals can range between 10 and 40 pounds. Adult weights of wolves grade from 35 to 135 pounds. The range of sizes among the different wild species in the canid family is almost the same as in the different breeds of dogs. Is a coyote just a stunted wolf, or the wolf an overgrown jackal? Is the wolf just a jackal adapted to life in a northern forest?

Think of all the possibilities for evolutionary size changes. Size can change just by speeding up, slowing down, lengthening or shortening the growth period. At some point during their life cycle, wolves begin to grow, and they stop growing at some later point. By changing the onset or offset of growth, the size of the animal can be changed dramatically. If I delay the onset of growth, or accelerate the offset, the resulting individual will be smaller because it grows for a shorter period. Or if I accelerate the onset and/or delay the offset, the individual will be larger than the ancestor. These biological mechanisms can easily be manipulated. Simply turn the growth switch on and off at different stages of development and the adult's final size is changed.

Changing the timing of the growth process seems to be exactly the difference among dogs (and breeds thereof), wolves, coyotes, and all species of jackals. Changes in the timing of the growth process is called heterochrony (from the Greek *hetero,* "changes," and *chronos,* "time").

All the "species" and/or breeds of the genus *Canis* have exactly the same gestation period of sixty-three days. They all bear neonates that are almost exactly the same size, shape, and brain volume at birth. They all have (practically) identical growth and developmental timing up to birth. At birth, border collie pups are the same size as Saint Bernards, wolves, coyotes, and jackals. The dogs will differ in color and have floppy ears, but all newborn puppies, regardless of breed and species, behave identically. They all have the same suckling responses, the same care-soliciting calls and alarm calls, and the same comfort-seeking behaviors. There are some differences in the size of newborn pups, but they are probably not genetic. The ability of the female to invest in the litter can affect the size. For example, some of my females have given birth to litters of thirteen or fourteen pups, all smaller than normal. The same female, with a litter of only eight pups, produced normal-sized pups. Toy breeds have small pups, but this is partly because they don't have the body mass to make the usual investment in pup size. They still have the official sixty-three-day gestation.

In other words, throughout the evolution of the genus *Canis* there has been little to no change in the embryology or gestation of the embryo. Dogs have been just about as conservative as all the other members of the genus in puppy development. Why has the genus been so conservative about puppies? All puppies are adapted to the same environment—the puppy size, shape, and behavior is an adaptation to a niche called Mom. When you think about it, puppies are very sophisticated, with highly specialized and adaptive structures. Puppies are perfectly designed for their niche, and there has been no tendency to change that design. The environment provided by the mother has remained stable for millions of years, and so in the true Darwinian sense, puppies are an exquisite adaptation to that environment.

For the most part, it is after birth that breed and species differences begin to appear. After birth the different breeds of dogs and the wild canids begin their metamorphosis from the highly specialized and functional puppy shapes to their niche-specialized adult shapes. The different breeds of dogs and species of wild types carry out that metamorphosis at different speeds. It is the same genes, operating at different speeds. Evolutionary biologist Richard Schneider and I studied the growth-rate differences between wolves and coyotes, and asked the question, why do coyotes have a different size and shape than wolves?

Coyotes and wolves are all the same size at birth, and at three months old they still are the same size. But after three months, the wolf growth rate continues to accelerate, while the coyote rate slows down. Coyotes start to grow their adult long nose right after birth, while wolves keep their puppy-shaped noses for almost six weeks. Wolf puppies' heads get bigger but do not start changing shape until the pups are four to six weeks old.

The differences between wolves and coyotes are not differences of genes per se. The same genes in each species are responsible for the growing noses, but they have different alleles (forms) of those genes, and those differing alleles are what skew the onset of growth and offset of growth in the two species. The differences in wolf and coyote head shapes are a result of when they start to grow, how fast they grow, and

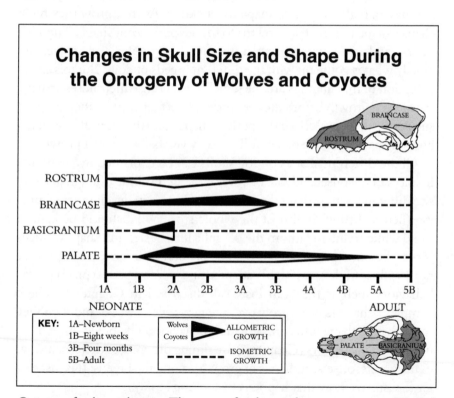

Ontogeny of wolves and coyotes. The noses of wolves and coyotes start growing and changing shape at different times after birth. They also grow and change at different rates. Coyotes and wolves have different-sized and -shaped heads because of these heterochronic differences in developmental timing. The huge differences in dog head shapes and sizes are also the results of different onsets, rates, and offsets of developmental events. (Schneider and Coppinger, unpublished manuscript)

when they stop growing. Therefore they end up with different-sized and -shaped heads. Note in the diagram that both wolf and coyote pups stop changing shape at the same time. Essentially, at three to four months, they have their adult-shaped heads, although they are small. At three or four months, coyote and wolf heads are still about the same size, but they are different shapes.

HOW TO CHANGE SHAPE

Can one change size without changing shape (isometric growth)? Puppies have a highly adaptive shape for suckling. As they grow they must change shape (allometric growth) to the adaptive adult shape. The allometric and isometric growth differences of coyotes and wolves are exactly what is going on with the different breeds of dogs. Borzois are an exception to the "all pups are the same" rule because their noses actually start rostral growth before they are born. (This may be why they are poor sucklers—they don't have the perfect shape at birth.) In bulldogs, nasal bone growth has a late onset, followed by very slow rates of growth.

All the different shapes of dog skulls can be explained by slight differences in the onset and rate of development, resulting often in just tiny size differences in the paired nasal bones. Change in developmental timing (heterochrony) is almost the whole story. What should be kept in mind while trying to fathom the complexities of size and shape changes is that the same processes apply to behavioral differences. Puppies are born with complex and highly adapted and stereotyped (typical) puppy behaviors such as the "I'm lost" call discussed in Chapter 6. These complex behaviors must grow over the next several weeks and then they must change shape into complex adult behaviors. The breed differences in adult dog behaviors (Chapter 6) are timing differences (heterochrony) in the onsets, offsets, and rates of growth among the breeds. It is just easier to visualize how that happens by looking at skulls.

There are limits to how much the timings can be changed. The extreme faces of the bulldog and the borzoi are perhaps the limits of what is possible. These breeds might not be able to survive any additional selection for shorter or longer faces. There are developmental restraints beyond which no animal can survive.

At first, one might say that the differences in nose lengths between borzois and bulldogs are genetic characteristics. But they are only genetic differences in the initiation and cessation of growth. There are no genes for face shape per se. There is no map in the genes that predetermines the shape. This is an important distinction in phrasing. I described border collie herding behavior as genetic, but said there were no genes for herding. Now I say that the growth of the dog's nose is genetically timed, but the shape of the dog's nose is not genetic, in the sense that there is a blueprint for it. Noses are the result of a number of developmental processes. To get technical for a moment, here is how the embryologist Pere Alberch said it.

> ... the role of development in evolutionary processes of morphological change is twofold. On one hand, the structure of the developmental program defines the realm of possible novelties ... while, on the other, the regulatory interactions occurring during ontogeny can accommodate genetic and environmental perturbations and result in the production of an integrated phenotype ...
> [Pere Alberch, 1982b.]

This statement will make many feel the way I do when someone quotes Shakespeare in an essay. I was always grateful when I studied Shakespeare when the professor would explain the meanings as he read and reread the otherwise opaque passages. What Alberch is saying (in dog talk) is, mainly, the role of development in changing the shape of each breed's nose (evolution) is twofold. On the one hand, there are timing constraints. For example, it is a constraint that the neonatal puppy nose must be an adaptive shape when the puppy is born or it can't suckle properly. The transformation of the puppy skull shape to the adult skull shape cannot begin until after the pup is suckling. That is a developmental constraint! Since building the adult skull cannot start until after birth, the total number of different adult skull shapes is limited.

The borzoi skull, however, is an exception. In order to get that really long nose and the ocular overlap, development must start before birth, which puts pups at risk. If the system were pushed just a little further, none of the puppies might be able to survive. That is the first developmental constraint. Indeed, if humans didn't give them extra support at birth, they might not survive. Why can't wolves develop more ocular overlap and have better depth perception? Because they cannot give long-faced pups the necessary support. That is one reason wolves have been so conservative over the six million years. Because there are important developmental constraints that cannot be avoided.

The second fold of Alberch's "role of development" incorporates the interactions that calibrate one growing bone's relationship with its neighbors. If the nasal bones are growing rapidly, the bones connected to them must also grow rapidly, especially in the areas of attachment to the nasals. Whatever the growth rate of the nasal bones, they have to fit with the rest of the dog's head. That means the head and all its parts have to accommodate to whatever the nasals are doing, to make an integrated skull. Realize that the integration and accommodation are reactions of the bones themselves, and not written in the genes.

Take the brain cavity as a simple example. The brain is surrounded by bone. The size of the braincase in a tiny puppy is tiny. In an adult dog it is big. As the puppy brain grows, the brain cavity of the skull must develop synchronously, every minute of every day. The skull and the brain must always be perfectly matched in size and shape.

In Chapter 4 I discussed the size of the brain and its dependence on the richness of the environment it which it grows. Because the environment varies for each growing puppy, there can be no accurate prediction as to what size and shape the adult brain will be. Brains grown in enriched environments are larger and have more connections than those brains grown in impoverished environments. Dogs that grow up in a visually enriched environment will have a different brain shape than dogs that grow up in an acoustically rich but a visually impoverished environment, even if they are identical twins. No two dogs can have exactly the same developmental environment, and no two dogs can have the same brain size.

But because each dog has a different-sized and -shaped brain, each dog has to have a customized braincase. The size and shape of the braincase could not be predetermined by genes. Imagine having one set of genes governing brain growth and another governing skull growth. They would soon be out of synchrony. We would have dogs running around whose skulls were too small for their brains. Imagine the headaches. Or, some dog's brains would rattle inside a too-large skull.

The simple point is that the brain determines the size and shape of the skull cavity. There are no genes for how big the skull is. The skull is an environmental response adapting to the growing brain. Skull shape and size are epigenetic, or above the genes. Shape and size are accommodations to the growing brain. This is how, as Pere Alberch wrote in 1982, " . . . regulatory interactions occurring during ontogeny can accommodate genetic and environmental perturbations and result in the production of an integrated phenotype." (*Ontogeny* is the history of an individual, and *phenotype* is what that individual looks like.)

Similarly, the eye sockets are an accommodation to the growing eyes. The brain changes shape in response to signals sent from the eyes. The brain cavity and the eye sockets constantly adapt to the developing brain and eye. The brain has to accommodate to the level of activity of the eyes. Thus all the parts of the skull are constantly adjusting to one another. The final size and shape of the skull are not predetermined by genes, but are epigenetic.

It may sound complicated, but conceptually it is very simple. As the skull grows, all the bones of the skull and all the organs within have to talk to one another. Each bone of the skull has to fit some space. Just as the brain has to fit into a cavity, so too do the nasal bones have to fit the nose. Like the brain, the nasal bones have to develop and shape the bones around them.

The borzoi and the bulldog are the extremes of the coyote and the wolf diagram. My student Abby Drake discovered that borzois are an exception to the rule, and start their rostral development before they are born. Bulldogs are more like wolves, starting nasal development late in the neonatal period, but unlike wolves they have a very slow rate of development. Since the bulldog's nasal bones don't appreciably lengthen, the palate and other bones accommodating the slow growth of these tiny nasals get pulled into weird positions. Weird here means that in an evolutionary (phylogenetic) sense, the palate has never occupied that position historically. In other words, the heterochronic shifting and retarding of nasal-bone development produces a head shape that is phylogenetically bizarre. Because the tiny nasals are slow growers, the palate is pulled up above the central axis of the skull, leaving the bulldog's teeth sticking out.

Now let us play some breed-development games. Bulldogs don't have genes for drooling—or do they? They have a mouth that doesn't close because of the upward angle of the palate. If I desired a breed of dog that drooled constantly, and in each generation I selected for those animals that drooled the best, I might not realize that I was really selecting for short nasal bones.

The borzoi's nose growth onsets early, and growth is so rapid it forces the palate and teeth forward and down below the axis of the skull, giving the dog that Roman nose appearance. Because the palate gets depressed, it pulls the eyes forward and closer together. Bringing the eyes forward and closer together gives the borzoi more ocular overlap. They should have better depth perception than other breeds. Do they chase rabbits by sight because they are better adapted to seeing three-dimensionally than other breeds? Do they have genes for seeing three-dimensionally? Yes and no! They have a longer nasal-growth period, timed at an earlier age, which puts their eyes in a phylogenetically bizarre position—one that results in more ocular overlap and better depth perception.

If I wanted to select for gazehounds that hunt rabbits using a refined depth perception, I would sort through my dogs and pick those that worked best, and breed the best to the best. I might not realize that all I am really doing is selecting for longer nasal bones. The eyes themselves are no better than in any other breed. They are just in a better position to see forward.

Now, could I develop a breed that sight-hunted rabbits and also drooled? Maybe not. That could be one of those developmental constraints.

The point is that the change in timing (heterochrony) of the growth of one small character (the nasals) can change the shape and the functioning (behavior) of many organs because each organ must accommodate to the changes of its neighbors. If one is selecting for a flat-faced breed for aesthetic reasons, one might not realize that the eyes are being pushed to the side of the head. Breeds like the bulldog have better peripheral vision, more like a robin. The flat-faced breeds, however, have no space to develop the turbinate bones under the short nasals. Thus, they don't have the normal amount of respiratory tissue that normally covers these bones. As a consequence they can't breathe properly. Even more important, they have less of the tissue that helps cool the brain. The flat-faced breeds often have chronically low oxygen tension in their blood, and are easily heat-distressed. Could we select for a flat-faced dog that had normal oxygen tension? Probably not. Could we develop a short-faced gazehound that had endurance? No—we would run into developmental constraints.

THE SHAPE OF INTELLIGENCE

Borzois can focus on game better than bulldogs, not because they have better eyes but because they have longer nasal bones. Now, let's add a couple of concepts together. Borzois are going to see the world differently as they grow up; this different view should develop a differently shaped brain than a bulldog's. Their brains, enriched in different ways, will create differently sized and shaped craniums. To judge an Afghan hound as less intelligent than a border collie by means of standardized measurements is to forget that the two breeds have very different developmental histories, even if they were raised in the same environment. They see the world from slightly different angles as they are growing up. They can't be going through the critical period of brain development in the same way. Therefore, on a standardized test, they will necessarily score differently.

Does that mean that their basic brains, or their basic intelligence, are genetically different, or does it simply mean that they grew up in a different developmental environment? They grew up in a different shape. Intelligence is always epigenetic. Intelligence, like bones and behavior, is an accommodation to the environment in which the trait is growing.

I hope what becomes clear here is that the nature of bone growth cannot ever be separated from the nurture of bone growth. For exactly the same reason, the nature of the behavior cannot be separated from the nurture of behavior. The resulting physical and behavioral conformation of a dog is not genetic in the sense that genes prescribed or predetermined the adult traits. Each and every characteristic is an epigenetic environmental adaptation or accommodation to the space in which the trait finds itself growing.

Now perhaps Alberch's passage on development is more understandable.

The behavioral differences among pointers and retrievers and border collies are results of heterochrony—the dissimilar developmental timings of eye-stalk-chase motor patterns. By selecting for timing changes in gene products (heterochrony), I can change the shape of a dog's behavior in exactly the same way I select for longer or shorter nasal bones. A main theme throughout this book is that behavior is as much a shape as a bone or an organ. When I discussed the outrun pattern of border collies, I noted that each dog had its individual shape of performance. I see that shape in the same way I see the shape and size of the brain cavity.

RAPID EVOLUTION OF BREEDS

Now I am back to an earlier problem. Shifted gene timing leads to the rapid evolution of both shape and behavior. The differences among the breeds are perhaps all due to slight changes in developmental timing. Here is the rub. To change the gene timings I must have variation in timing in the ancestors in order to select for any differences. Within breeds or species there is often little variation. When we say that they breed "true," the implication is there is little genetic variation. If they breed "absolutely true," it would mean there was no genetic variation at all. No evolutionary change could occur by natural selection (or artificial).

If there was even slight variation in some characteristic, I could select, in a Darwinian sense, for change. If I wanted to select for better health in my bulldogs, I would breed the healthy, more robust creatures together and gradually (eventually) create a healthy strain. This is the system Belyaev used when he bred those foxes with short flight distances to other foxes with short flight distances and produced, within a few generations, tame foxes. The more variation in a character, the faster I could select for change of that character.

Gradualism by artificial selection is slow. It ought to be obvious that natural or artificial selection cannot be the way our modern breeds were created. Breed development happens too fast to be a result of gradual selection for single characters. The record on breed development is reasonably clear that most breeds were created by hybridization. One can create variation, or shifts of timing in gene products, by hybridization (aka crossbreeding, mongrelization).

Hybridization is a way of instantly producing novelty. As a matter of record, of the two hundred to four hundred breeds (pick your favorite number), the majority have been created since the nineteenth century by crossbreeding. The history of a breed frequently contains the assertion that this is an ancient breed, and then goes on to admit the breed was rejuvenated in the 1920s, say, by crossing it with something else. The golden retriever started with an unusually colored (mutant) dog of superior behavior that was then bred to Irish water spaniels, setters, and so on, to "perfect" the breed. The German shepherd, the Doberman, and the rottweiler all have documented histories of which breeds went into producing the modern breed. Bedlington terriers, which are so popular in crossbreeding with greyhounds to produce lurchers, are themselves a cross of greyhounds and terrier. The sled dogs, which are so perfect at marathon distances, are a meltdown of any number of breeds brought to the fields of the gold rush, put to work, and the best performers got bred together no matter what their breed.

Why aren't mutations the source of variation? Simply because mutations are tiny, somewhat insignificant changes. The mutation for golden color is not what made the dog a great hunter. Other mutations produce radical and unpredictable changes, such as the achondroplasia of dachshunds and basset hounds. These are saltations that might then be adapted and selected for some new task. But short legs were not selected for in the Darwinian sense. Mutations are usually little saltations of form such as coat color, but sometimes are large saltations like change in leg length. There are two things to remember about this: the mutation is a saltation with unpredictable results, and the initial result was not selected for.

Biologists have known about hybridization and the power of it in evolutionary change for a long time. As Darwin himself suggested, the cross between wolves and jackals might have produced the necessary variation for the evolution of the first dogs.

Hybridization is not an entirely popular subject among evolutionary biologists, partly because it raises that sticky question of speciation. How can you have crossbreeding between sexually isolated animals? But occasionally one of the great evolutionary biologists will mention the potential power of crossbreeding two different forms. By the mid–twentieth century a number of the heroes of evolutionary theory had considered hybridization a natural source of genetic variation.

Ledyard Stebbens, often given credit as one of the great synthesizers of the Darwinian theory, noted that new forms or saltations could be produced instantly by hybridization. (Here is the word "saltation" again—leaps of form, not gradualism.)

In 1930, J. B. S. Haldane, thinking about evolution, wrote, "[there is] every reason to believe that new species may arise quite suddenly, sometimes by hybridization, sometimes perhaps by other means. Such species do not arise, as Darwin thought, by natural selection. When they have arisen they must justify their existence before the tribunal of natural selection."

I would like to reword that to "new breeds of dogs arise quite suddenly, often by hybridization, producing bizarre shapes to justify their existence before the tribunal of natural or *artificial* selection."

In the 1930s and 1940s, the most famous and infamous experiments on hybridization, on dogs, were done by Charles R. Stockard at Cornell University. Stockard was an embryologist who had a wonderful history of designing unusual experiments that asked questions about animal development. I always use his fish experiments in teaching the critical period hypothesis. By cold-shocking fish embryos at a precise time in development (critical period), Stockard could produce adult fish that were cycloptic. There are no genes for two eyes. Two eyes is a developmental response to an environment, just as his one-eyed fish were the response of the same set of genes to a different environment. (If one could just get puppy raisers to see that environmental inputs at critical periods have the same kind of lasting effects as those cold shocks!)

In the dog hybridization experiments, Stockard crossbred several breeds of dogs with one another to see what would happen. He bred the tall ones to the short ones, the flat-faced to the long-faced. He produced not only a huge variety of forms by crossbreeding, but he got shapes that were bizarre and unpredicted. He got new shapes that were not averages of the parents. He labeled them "structural disharmonies." Structural disharmonies might be monsters. These were animals where the parts did not fit together in a functional way. The back legs were longer than the front legs. The lower jaw was longer than the upper jaw. The accommodation between the growing bones was stretched to the limit. But structural disharmonies also meant something else. These were new and unexpected shapes.

At the beginning of this chapter I asked the question, Why do we have so many different breeds? Why, if the dog's shape and behavior are so malleable, is the ancestral wolf so conservative? Why does the wolf hang around for millions of years and change size a little, but not much else?

Haldane says change in form could arise because of hybridization, but then the new and novel result of hybridization must be tested before the tribunal of natural selection. One answer to the conservative-wolf question is that those dog-breed shapes potentially available to wolves couldn't stand the test of natural selection in the wild. Maybe there have been wolf pups born with borzoi faces that just died. Developing a better ocular overlap might be an adaptive advantage for an adult wolf and a disadvantage for a puppy. It would be a developmental constraint for wolves if, because of the changes in the timing of nose growth, the wild-born puppies were inferior sucklers.

NEOTENY, PAEDOMORPHISM, AND THE EVOLUTION OF DOGS FROM WOLVES

As I've just shown, the way to create change from one dog breed to its descendants is to alter the timing of gene products (heterochrony). The easiest and quickest way to maximize variations in timing is to hybridize different sizes or shapes of dogs, creating new and bizarre forms within one generation. Then, by selecting among these new forms, new breeds are created and standardized.

But let us explore how heterochronic changes could alter a wolflike creature into a doglike creature. In the first chapter I noted that the basic diagnostic doglike features were adaptations to a new niche. Those adaptations were changes in size and shape. The overall size of the dog on Pemba was reduced from wolf (perhaps forty-five pounds) to dog size (twenty-five pounds). Such a reduction of the overall size is easy just by changing the timing of growth.

But size is not the diagnostic character that identifies dogs. Their proportionately smaller heads, teeth, and brains are the traits of the generic dog. I can't get those features just by speeding up or slowing down overall growth. The head growth has to be disproportionately slowed, or retarded, in relationship to body growth. And it turns out that this is fairly easy to accomplish, heterochronically speaking.

Look again at the chart of wolf- and coyote-skull growth on page 301 and note that both have their adult-shaped heads at about four months. But their heads don't get their adult size until they are seven or eight months, or even a year or more in wolves. What appears to happen with dogs is they get their adult head shapes at four months just like everybody else, but their heads' growth rate slows at that point, compared with the growth of the rest of their bodies.

That means that adult dogs have head sizes that look more like that of a four-month-old wolf puppy. It is interesting that my hundred-pound livestock-guarding dogs have the brain size of a four-month-old wolf puppy. In Chapter 1, I constructed a hypothesis where natural selection favored dogs with heads and brains smaller than those of their adult ancestors. That could happen by the retarding of the growth of the head at some stage of wolf-puppy development—an example of neoteny, or a retention of a juvenile character into the adult form.

In Chapter 1, I also hypothesized that dog behavior was an adaptation to feeding in the presence of humans, and soliciting food from humans. It is interesting to note that four-month-old wolf puppies with small, adult-shaped heads and four-month-old brain sizes are also in the stage of development when pups are soliciting care and food from adults.

The neoteny theory describes a heterochronic process whereby dogs developed dog shapes and behaviors by retaining wolf juvenile shapes and care-soliciting behaviors longer into adulthood. Dogs, according to the theory, have physical traits characteristic of wolf puppies rather than wolf adults. Dogs also, then, have behaviors of wolf puppies, not wolf adults. Wolf puppies, for example, sit dutifully at a rendezvous location waiting for the pack to return with food. Does that mean that every time I come home to my dog, who sleeps on the front porch, seldom straying more than a few feet away, and then jumping all over me in anticipation of dinner, I think he is acting like a wolf—a wolf puppy?

Many authors, including myself, have suggested that dogs are an example of this kind of heterochronic, retarded neotenic development. Other researchers think some breeds of dogs are paedomorphic, not neotenic. The distinction may seem slim, but it is biologically significant. Paedomorphism is a *result,* a truncation of development, where the animal becomes reproductive in an ancestor's juvenile stage.

These are all intriguing speculations about how dogs evolved—from wolves or from each other. But the first danger here is that one must be careful to avoid the sloppy thinking that dog behavior is "genetic." After all, Brian Plummer's King Charles spaniels were perfectly good hunters. The second danger is that one has to be careful to consider whether tendencies toward juvenile motor patterns are homologs, meaning they were directly selected from the wolf ancestor thousands of years ago. It is more likely that modern dog characteristics were inherited from other dogs.

Most of the paedomorphosis or neoteny theories came from the perception that dogs tended toward short faces, when compared with wolves. Adult dogs looked like puppy wolves. I fell into this trap myself during the early years of our sheepdog project. But adult dogs don't have puppy-shaped wolf heads. Robert Wayne demonstrated this conclusively in his 1986 paper on the cranial morphology of wild and domestic animals. He found that all breeds of dogs have identical skull length proportions to wolves and all the other wild canids. It seems impossible, but adult bulldogs, adult borzois, and adult wolves have identical ratios of palate to total skull length. And neonatal dogs and neonatal wolves and coyotes also have identical ratios of palate to total skull length. But the length ratios between the adults and the puppies, as you might expect, are different. If dogs were paedomorphic wolves, then they would have to have adult ratios similar to wolf neonate or puppy ratios. And they don't. Dogs' heads look short because they are wider (in some cases) than wolves' heads. Wider is not a paedomorphic trait.

Since all breeds of dogs lose their puppy teeth and produce adult teeth, truncation of growth and development would have to be at the age when wolves had their second teeth, which is at four to six months of age. Because of these adult anatomical traits of all breeds of dogs, it is virtually impossible to argue successfully that any breed is globally paedomorphic. Could the argument be made that some breeds are neotenic results of others? Sure, but given the tremendous hybridization that has gone on between the breeds, it is impossible to ever discern it. Golden retrievers, for example, are neotenic forms of which of the several different ancestral breeds that were hybridized to create them?

I had created a chart of what I then considered as the more and less neotenic breed profiles, published in *Smithsonian* magazine in 1982. Since then, other researchers have produced a tremendous new literature on heterochrony, and I spent several years measuring dog skulls and dog behavior in detail. Using marvelous new computerized skull-measuring instruments and new methods of statistical analysis, my student Richard Schneider and I spent months in the catacombs of the Smithsonian vertebrate collections. It didn't take too long for us to give up neoteny as a useful concept for describing dog evolution. That doesn't mean that the neoteny hypothesis is wrong, but rather that nobody has yet found a method of testing it.

There are huge developmental differences between dogs and wolves. If you don't socialize a wolf to humans before it is nineteen days old, it is too late. In dogs, it is too late at eight, nine, or maybe ten weeks, depending on the breed. What happens if you crossbreed two dogs, one which has a six-week limit and another with a ten-week limit? Are the offspring average or are they novel, bizarre, and uniquely different? In Chapters 4 and 6 I explained the enormous differences in breed-specific behavior between livestock-guarding dogs and herding dogs. Guarding dogs have the onset of predatory motor patterns *after* the social window has offset, and border collies have the onset of eye-stalk behavior *during* the socialization period. In wolves, fear onsets before they develop their juvenile food-begging behaviors.

Which of these canines is more neotenic or paedomorphic? Border collies vary from wolves in their much earlier onset of eye-stalk adult behaviors. They can accommodate social eye-stalk into their play. The presence of eye-stalk produces an integrated behavioral phenotype. That integrated behavioral phenotype changes the growth shape of the brain, which changes the shape of the skull. Is the difference in the skull shape of border collies, livestock-guarding dogs, and wolves genetic? Is it neotenic? Is it paedomorphic? None of the above. Border collies are phylogenetically bizarre. And each border collie is an epigenetic response to incorporating eye-stalk games in the social behavior at some critical stage of brain development. Each border collie is individually bizarre, in that there has never been another dog like it. How a dog looks, then, is intricately tied to how it behaves, from its molecular to its holistic levels.

That is why dogs look the way they do. The joy for me of studying all the different forms of dogs, and working with literally thousands of dogs, no two of which are ever alike, is simply delighting in the continuous variation they show.

Throughout history, the people who knew dogs and created breeds did so because they wanted to produce something new and better. Even now, the lurcher people of Europe are looking for the magic combination that produces a superior mink-hunting dog. And those of us who are trying to win a dog race, or make a better varmint dog, know that variation engenders variation. From the new combinations, new tasks can be achieved. And there you have it. All the various variations people have developed are capable of producing a cascade of new variations. The production of numerous and bizarre forms increases the possibility that we can find and produce new breeds to work in diverse environments, in new ways, with people.

The possibilities of creating new breeds that perform unusual tasks are almost limitless. Theoretically, therefore, we should be able to find new and different symbiotic relationships with dogs. Maybe we could change some of the less desirable relations we have now, such as amensalism with household dogs and dulosis with service dogs. Perhaps if we tried to understand the nature of dogs and their evolutionarily bizarre abilities a little better, we could produce perfect ones without jeopardizing their genetic health. The creation of breeds by cross-breeding has been a remarkable opportunity for humans to exert forces of selection on a species of canid, and for the most part we achieve healthy results. The evolution has been a bonanza for dogs. This is evident not only in how many breeds and dogs there are, but in how much direct care they get from humans.

But as I pointed out in the household-dog chapter, dogs are also at great risk in the wrong hands. Many dog breeders have produced a contrary effect in trying to preserve breeds. They treat breeds as if they were species, and sexually isolate small populations of them in an attempt to preserve their historic, ideal phenotype. Sexual isolation from the greater population of dogs leads almost inevitably to dire consequences for those dogs that get trapped in a pure breed. Indeed, the idea of trying to modify a breed's behavior into a more tractable type of pet, while holding its form constant, seems not to work very well. Holding the size and shape constant while changing the behavior might well be one of those developmental constraints that don't work, like trying to get ocular overlap and robust drooling in the same animal.

Another severe problem with locking dogs up reproductively is the problem of inbreeding. Once the stud book is closed on a breed, it is unbelievable how fast they become inbred. I was sitting in a review session at The Seeing Eye in New Jersey one afternoon with John Pollak, a geneticist from Cornell, and I asked how fast inbreeding will occur once a population is isolated. A true teacher, he led me through a little exercise.

How many founding sires do you start with? If you have just one, then all the first generation will be siblings or half-siblings. By the second generation, all breedings are inbreedings. If there are two founding sires (unrelated), then the third generation is inbred. So he developed a formula for me to go home and practice with. If I started with five hundred unrelated founding males when I closed the stud book, then by the tenth generation I will start inbreeding. That could be only fifteen years after the stud book was closed.

If I created a breed of dogs in 1900 (that is, closed the stud book) with five hundred males, currently that breed would have been inbreeding for eighty-five years. They are caught in a genetic trap. And what can possibly be done about it? Open the stud book.

The pure breed story is worse than that. Starting with five hundred males, I get ten good breeding years if I use all the males equally. If the members of the breed club begin to breed only to the champions, then the inbreeding is accelerated. If the stud book closed on five hundred males but every female is bred to this year's grand champion, then inbreeding starts next year. Is it such a wonder, then, that our purebred dogs have so many breed-specific diseases, increasing all the time? Consider the advice of the experts who counsel breeders to eliminate from their breeding programs those dogs that exhibit retinal atrophy or hip dysplasia. The inbreeding coefficient increases more rapidly. The breed is in big trouble.

The old-fashioned breeders who continue to create dogs by cross-breeding for specific, specialized tasks, like the lurcher breeders of Europe or the sled dog drivers, are, by and large, disdained by pure breeders. I have been chastised many times by newcomers to the world of the uncommon guardian breeds. How could I possibly crossbreed the pure white Maremmano-Abruzzese with those gray-and-black Šarplaninac? Well, I say, in the first place, my understanding of the transhumance leads me to believe that the Maremmano-Abruzzese and the Šarplaninac are not pure breeds at all. And in the second place, improvement of plants and animals, when performance is the goal, relies on crossbreeding and hybridization. The ability of agriculture to produce the quantity and quality of animals and plants it does depends heavily on crossbreeding and hybridization. The successful techniques of crossbreeders of working dogs are practically unheard of outside of their fields. What the purebred breeders forget is that golden retrievers and every other modern breed are products, originally, of crossbreeding. That is why they have been good dogs. At least in the beginning, they had the health and energy that are known as hybrid vigor.

Surely we owe dogs more than tightly restricted lives and distorted body shapes. Surely we can give up the eugenics of the pure, the perfect dog, and create instead a population of well-adapted, healthy pet dogs. In my wildest dream, I imagine people who have given up the "What kind of dog should I get?" question and gone to "I would like to make a dog for this task."

CHAPTER 12

Conclusion

Perhaps my intent in this book was not to answer all the questions about dog behavior, but rather to raise some. Dogs, maybe more than any other species, are surrounded by myths, fictions, and factoids. Lassie comes home, like Ulysses, performing heroic feats of incredible agility. Lassie recognizes good people from bad, has unbelievable senses of hearing, sight, and smell, and a cognitive ability barely exceeded by humans themselves.

Mythology caricatures dogs. It conjures images that are fun and inspirational, and there is nothing wrong with stories about heroic dogs unless they interfere with dogs' health and well-being. However, a better understanding of dogs' true behavior, their intentions and their motivations, must lead to a deeper, richer interspecific relationship between people and dogs. The village dog is a far greater phenomenon than the cartoon Lassie. Since its origin, the village dog has adapted to a staggering array of tasks set for it by humans. Like the metamorphic shape-shifters of fairy tales, the original animal has transformed into a myriad of shapes, sizes, and behaviors. Dogs rival people in their abilities to adjust to a variety of circumstances. Unlike people, they can make genetic adjustments at a rate that challenges our theories of transmutation.

Recently I helped a film crew make a documentary film about dogs for the PBS science program *Nova*. The producers wanted scenes with village dogs, but didn't want to go to Africa to get them. The first rule, then, I suggested, was to go south. The nearer to the equator, the larger and more typical the population of village dogs. Second, just about anywhere there was a permanent settlement of humans with open dumps, I was sure we would find the desired dogs. That was my confident advice.

The crew went ahead and made arrangements to shoot the footage at the Tijuana dump just over the border in Mexico. Now I began to worry. I worried that this might not be the best dump to illustrate the origins of dogs. Tijuana is a fast-growing, big city, and might be described as a boomtown. It is not old and stable, the way I think of ancient villages where dogs originated. Given the history of trade agreements between the United States and Mexico, the creation of factories and worker housing, and an immigrant population of people and dogs to the city, I was afraid that Tijuana would not be illustrative of much of the history of a developing dog-people symbiosis. In Massachusetts, I worried that Tijuana wouldn't have that natural feel to it.

Well, I was right and wrong at the same time. However wrong I was I spent several of the most exciting field days of my life with the Tijuana dump dogs. Instead of being anomalous, the Tijuana dump was strikingly representative of the many relationships between people and dogs. It's probably not good for my image, but I have to say that the dump at Tijuana was one of the most fascinating places I have ever studied dogs.

First impression: The Tijuana dump is beautiful. Second impression: It is like a military operation. It is at the top of a small mountain. A convoy of trucks comes to the top of the mountain and dumps the refuse of a developing city. Then, big equipment pushes the accumulating debris over the top toward a valley below. Other equipment moves mountains of dirt to cover the wastes. At intervals, standpipes are inserted into the filled spaces to allow the biogas of the rotting materials below to escape. The wind blows across these vents, creating an eerie music. When the biogas envelops you, you sense that sinking feeling of doom.

In the evening, as the sun sets, it shines across a myriad of multicolored objects such as plastic shopping bags, throwing a psychedelic effect over the scene. Everywhere are patches of different colors. As I looked directly into the setting sun, there were people and trucks silhouetted, along with thousands and thousands of seagulls, against the sky.

And then I saw the dogs. Every one was peculiarly backlit, as if it moved within its own halo. Every dog and its shimmering shroud explored the contents of the colored plastic bags.

I began to pick out dogs everywhere, and guessed there were hundreds of them. The dogs paid no attention to the gulls, which I think is significant. Here are two very different species that have invaded the waste resources of people. For the gulls, this may be close to a primordial event, and maybe in ten thousand years they will have doglike behaviors, living in obligatory symbiotic relationship with people.

This dump is perfect for filming a fantasy about the Mesolithic origin of dogs. The Tijuana dump might have ecological components of all the world's dumps. It might be called Everydump. Of course, the first dumps, eons ago, couldn't have had the complexity of the Tijuana dump. Not only does the Tijuana dump contain a greater variety of waste products than we would expect in the Mesolithic village, but in this dump are layers of social behaviors of both dogs and humans. The social interactions between people, between dogs, and between people and dogs are a most complicated caste system. The range of the castes of both dogs and people is from the pristine to the unbelievably unclean.

The first layer at the dump are the people. The truck drivers are the constantly moving caravan traders. They stop at the tollhouse and pay the $18-per-ton fee and then zoom to the dumping zone via the road that winds through piles of debris. After their wares are deposited, they return to their suppliers for another load. These merchants of trash continually supply this center with the necessary resources required to support the thousands of lives of the occupants.

Occasionally the drivers stop and have lunch, coffee, or a smoke with other drivers. They don't necessarily know one another but their common goal bonds them together for an occasional social interaction. And perhaps, like me, they pause to enjoy the view.

The employees of the dump are definitely in charge. Everything that happens there, including our filming, is done with their permission. Each employee has a rank and is spaced out according to rank. There are dwellings for the managers, and messengers scurry back and forth to the equipment operators, mechanics, or biogas engineers. There are generals, lieutenants, aides, and enlisted men. They are ever busy. Their pace is almost frenetic as they try to keep ahead of the convoy of trucks. They are more social with one another than they are with other occupants of the dump. They talk and interact in various ways during the day, and I suppose they go home and have dinner with their families.

A second layer of people immigrate to the dump daily. These are the miners. Some collect rags, others collect metal or other saleable objects. They seem to be specialists. Each has a searching image of what he wants. Each person is able to package his findings in a marketable way. Rags, for example, are consolidated in bundles of specific sizes. These bundles are then carried or trucked (in small trucks) back out of the dump.

Unlike dump workers and dump truck drivers, these resource collectors are competitive with one another. There are a limited number of rags in the dump. To make a decent living, one has to have a technique for acquiring enough to make a living. Some appear to be territorial, others employ labor, like a wife. Some arrive early and can identify the rag trucks, and so they get first pick. As they work, they are not as social as the official workers because of the spacing, but they call to one another, and they often meet and sit in discarded chairs for a smoke.

A third layer is the people who live in the dump. Sometimes they live alone, and sometimes they share a shelter with someone. The shelter might be a big cardboard box, often furnished. Clothes and other possessions obviously come from the dump, and some of them are quite nice, even new. These people share things, campfires for example. They, like the miners, are competitive for resources and tend to be spaced out in ways that reflect social and resource needs.

And, if all be known, this group of people is probably competitive with the dogs for food.

The symbiotic relationship of dogs with people is obligatory. Further, the dogs of the Tijuana dump appear to honor a social distribution that reflects the social relationship of the people.

There are the unclean dogs, whose permanent home is in the dump. They search carefully for food. They have a favorite place for sleeping. They often share their sleeping places with other dogs, but they tend to feed alone. Unknown people from far away send their refuse to the dump and these dogs find the edible parts and make a living. They have no other contact with people. Like the gulls, they feed among people, but people are only part of the background. Like the gulls, they never socialize with people, and at best they move away if people get too close. This is the classic commensal relationship of dogs to people. They are dependent on people, but they don't know it, and they don't benefit people in any way.

These commensal scavengers reproduce in the dump. Sometimes the pups are adopted by the people who live in the dump. I saw two tiny pups tied with some discarded string to a cardboard house. The occupant of the house had "gone to work." These pups would be fed and cared for by people during their critical period for social development. Perhaps

The Tijuana dump: unclean class. This dog of nondescript parentage lives full-time in the dump. Every dog is a direct descendant from the original dog, but this one is living the commensal life of the original dog.

when (or if) they grow up they will search for food with their human social companion, or guard the cardboard house. Maybe they will enjoy a mutualism with people. The dogs could use their superior noses to find half-buried food, while the people can use their superior intelligence and opposable thumbs and tools to extract it from the burial places.

Then there are social groups of dogs that immigrate into the dump each day from surrounding communities. Some would come in on the trucks. I would see a dog get out of a truck for the briefest of moments to sniff noses with another dump dog. But most of the migrants walk in. Some of these dogs migrate on a daily basis in and out of the dump with their miner friends. Some ragpicker dogs remain entirely in the dump and join their human companion anew, each day. I suppose there are specialized dogs that solicit food and companionship from only the ragpickers, while others concentrate on the metal collectors.

Some of these daily migrants don't seem to be very attached to people. They don't have collars, for example, and are the size and shape of the classic village dog. Their general good health suggests that they make most of their living in the nearby villages and come to the dump to supplement their food income. Maybe they just come to the dump for reproductive reasons.

This is certainly true of the other immigrant dogs. These are the working-pet dogs from the surrounding communities. They are the carpet-baggers and generally have nice collars. The leather-collar class. They all look like they might have some pure-breeding behind them. Some look like pure rottweilers, and I saw many classic pit bulls.

In the city and the surrounding bedroom villages, rotties and pit bulls are common. In Tijuana, many people keep these dogs for protection and entertainment. Betting on dog fights is popular. As disagreeable as this may seem, people claim benefit just as a pet owner would claim satisfaction from owning a good dog. The successful fighting dog can produce for the owner a cash prize and the increased sale of puppies from a superior individual. The dogs get cared for and the survivors get to leave their superior fighting genes to the next generation.

These immigrants of high breeding and independent food wealth

The Tijuana dump: leather-collar class. Feeding isn't the problem for these dogs. It would be interesting to know what their problem is. Why are they such meanies? I didn't like them.

coming to the dump have a very different demeanor. They are not opposed to feeding on the garbage. And they will fight over it. They are belligerent tough guys somewhat like Pemba's Chake Chake dogs in Chapter 2. They form into bigger groups, and, to tell the truth, at times I was a little afraid of them. More than once I got snarled at when my picture-taking activity brought me too close. And when I took the hint and backed off, they pursued me in what I thought was a threatening way.

This is very different behavior from that of the standard village dog, who will show its teeth on occasion, but whose typical response is to move slowly and steadily away from an intruder. Also, pure village dogs don't, as a rule, group together. After all, pure village dogs are competitive with one another. But these well-fed immigrant thugs are not competitive for food except in a ritualistic sense, although they are very competitive for social access. They can afford to waste energy in social play, even escalating it to open warfare.

By the way, I do not mean to imply that the aggression has anything to do with pit bull or rottweiler breeding. I've owned pit bulls, and I spent a day fishing with the nicest, sweetest rottweiler. These dogs are products of their developmental environment, as are, I assume, the people of this dump. Like most of us, dogs have very little choice about their developmental environment. And we as dog lovers have very little understanding about the parameters of that developmental environment.

Dogs might be evolving here at the Tijuana dump, but they could not have evolved here from wolves. All the dogs in this dump are products of dogs, and must have strayed in here at one time or another from some other ecosystem. But we might ask some questions about the societies of dogs found here.

How many generations of dogs would have to be born and live here before they are not regarded as strays anymore? If there is a population that bred in the dump for eight generations and has been sexually isolated (somehow!) from surrounding dogs, would they be purebred dump dogs? Would there be a nonrandom distribution of genes, signifying a distinct breed? Would the breed be a result of natural selection, or founder principle?

Could the commensal scavengers that depend on the Tijuana dump ever be purebred dump dogs in the sense of the natural breeds of Chapter 3? It is really doubtful. The natural breed that lives on Pemba lives on an island, where immigration and emigration are very low. The dog genes of Pemba cannot be added to or subtracted from except by mutation, by the tribunal of natural selection, or by random drift caused by natural catastrophes.

But Tijuana has stratified societies of dogs. Each layer is exposed to a different set of selective pressures. From a food and human-care point of view, these dogs belong to different social classes. However, they are only stratified in their food resources, and perhaps their hazard-avoidance behaviors. They all have reproductive opportunities across the ecological layers. The sporting-pet dog genes will escape back to the village-dump dogs. The village dogs will be periodically captured and their genes will appear as sporting-pet dogs.

A purebred dump dog is going to be bred by those swaggering pure-bred dogs from outside the dump niche. The dump dogs will also mate with one another. Since there can be more than one father to a litter, a purebred dump pup could have a sibling that is a hybrid between two ecotypes. Either type of offspring could be captured and raised as a walking hound or a livestock-guarding dog, or a pit dog. Depending on its success, any dog could become the founding sire or dam of a new breed of dog.

But breed of dog for what? If I took any pup out of the Tijuana dump and socialized it properly, what would it turn out to be? Could it be another once-in-a-lifetime dog? What jobs could it do? I think that any random pup coming out of the Tijuana dump has a good chance of becoming a good livestock-guarding dog, or a good walking hound. I know I could teach one to pull a sled, and herd sheep, and chase a ball, and be a hearing-ear dog. I bet my pup would be a fun dog, and completely different from the other dogs in my neighborhood. I bet it would be a picture of genetic health. After all, it has survived the most primitive of doggy conditions.

What couldn't my pup do? It probably couldn't be on a winning dog team, or win a sheepdog trial, or be best of show.

The time has come when we should give up our master-race view of which breed is best, or which dog is best. This modern emphasis has trapped our breeds in genetic isolation. I'd like to see our purebred dogs escape from these restraints. That doesn't mean I would mix them all together and go back to the original village dogs. It means simply that value should be placed on what the dogs can do, how healthy they are, and how they feel.

The real value of the pet dog, the companion dog, and the service dog is their behavior. My grandmother always said that handsome is as handsome does. It's time to realize the emphasis there should be on the "does." And also to take to heart what we often dismiss as sort of a joke: "A good dog can't be a bad color," and "I'd rather have an inch of dog than a mile of papers."

Good dogs are made after they are born, not before. No dog has the genes of a good pet. Certain breeds can make bad pets no matter what you do. We should recognize that and not try to make pets out of them. These are the breeds that display specialized sequences of motor patterns that are inappropriate around the home. Dogs with eye-stalk-chase behaviors, for example, do not make good pets or good service dogs because they are so easily distracted by stimuli that release the innate motor patterns (Chapter 6). It is a great mistake to buy such a dog as a household companion. Both dog and novice owner are likely to be very unhappy.

For most dogs, attention must be paid to developmental events during the critical period of socialization (one to eight weeks, and then eight to sixteen weeks). What happens during this time is crucial for making a good dog. Many people don't realize how much time that takes.

And while you are spending all that fun time socializing and training your dog, don't ever, ever treat it as if it were a wolf. Write it on the wall: "To be descended from a wolf is not to be a wolf."

And under that, write the great advice attributed to Canadian sled dogger Emile Martel: "Don't forget, they are only dogs."

Bibliography

The following books and papers were either consulted or referred to in the text. Not all of them are still in print, but are usually available in libraries.

INTRODUCTION

Bekoff, Marc, and Colin Allen. 1992. Essay on contemporary issues in ethology. *Ethology* 91:1–16.

Coppinger, Raymond P., and Charles Kay Smith. 1990. A model for understanding the evolution of mammalian behavior. In *Current Mammalogy,* edited by H. Genoways. New York: Plenum Press.

Fentress, J. C., and P. McCloud. 1986. Motor patterns in development. In E. M. Blass (ed.), *Handbook of Behavioral Neurobiology, vol.8: Developmental Processes in Psychobiology and Neurobiology.* New York: Plenum Press: 35–60.

Fox, Michael W. 1963. *Canine Behavior.* Springfield, Ill.: Charles C. Thomas.

———. 1972. *Understanding Your Dog.* New York: Coward, McCann and Geoghegan.

Gould, James L. 1982. Ethology: *The Mechanisms and Evolution of Behavior.* New York: W. W. Norton.

Grandin, Temple, ed. 1998. *Genetics and the Behavior of Domestic Animals.* San Diego, California: Academic Press.

Grier, James W., and Theodore Burk. 1992. *Biology of Animal Behavior.* Second edition. St. Louis, Mo.: Mosby Year Book.

Lorenz, Konrad. 1954. *Man Meets Dog.* Boston: Houghton Mifflin Company.

Ollivant, Alfred. 1924. *Bob, Son of Battle.* Garden City N.Y.: Doubleday Page.

Pryor, Karen. 1985. *Don't Shoot the Dog!* New York: Bantam Books.

Schneirla, T. C. 1966. Behavioral development and comparative psychology. *The Quarterly Review of Biology* 41 (3):283–302.

Scott, John Paul, and John L. Fuller. 1965. *Genetics and the Social Behavior of the Dog.* Chicago: The University of Chicago Press.

Serpell, James. 1995. Introduction to The Domestic Dog. In *The Domestic Dog: Its Evolution, Behaviour, and Interactions with People,* edited by J. Serpell. Cambridge, UK: Cambridge University Press.

Zimen, Erik, and Luigi Boitani. 1979. Status of the wolf in Europe and the possibilities of conservation and reintroduction. In *The Behavior and Ecology of Wolves,* edited by E. Klinghammer. New York: Garland STPM Press.

ONE

Belyaev, D. K. 1979. Destabilizing selection as a factor in domestication. *Journal of Heredity* 70:301–8.

Belyaev, D. K., I .Z. Plyusnina, and L. N. Trut. 1984/85. Domestication in the silver fox (Vulpes fulvus desm): changes in physiological boundaries of the sensitive period of primary socialization. *Applied Animal Behaviour Science* 13:359–70.

Belyaev, D. K., and L. N. Trut. 1975. Some genetic and endocrine effects of selection for domestication in silver foxes. In *The Wild Canids,* edited by M. W. Fox. New York: Van Nostrand Reinhold.

Brisbin, I. L. Jr. 1976. The domestication of the dog. *Purebred Dogs: American Kennel Club Gazette* 93:22–29.

Coppinger, Raymond, and Charles K. Smith. 1983. The domestication of evolution. *Environmental Conservation* 10:283–92.

Clutton–Brock, Juliet. 1995. Origins of the dog: Domestication and early history. In *The Domestic Dog: Its Evolution, Behaviour, and Interactions with People,* edited by J. Serpell. Cambridge, UK: Cambridge University Press.

Corbett, Laurie. 1985. Morphological comparisons of Australian and Thai dingoes: A reappraisal of dingo status, distribution and ancestry. *Proceedings of the Ecology Society of Australia* 13:277–91.

Crisler, Lois. 1958. *Arctic Wild.* New York: Harper and Row.

Darwin, Charles. 1903. *The Origin of Species.* Facsimile of first edition (1859–60). London: Watts.

Davis, Simon J. M., and François R. Valla. 1978. Evidence for the domestication of the dog 12,000 years ago in the Natufian of Israel. *Nature* 276:608–10.

Fentress, John C. 1967. Observations on the behavioral development of a hand-reared male timber wolf. *American Zoologist* 7:339–51.

———. 1973. Specific and nonspecific factors in the causation of behavior. In *Perspectives in Ethology,* edited by P. P. G. Bateson and P. H. Klopfer. New York and London: Plenum Press.

Frank, Harry, and Martha Gialdini Frank. 1982. Comparison of problem solving performance in six–week–old wolves and dogs. *Animal Behaviour* 30:95–98.

Frank, Harry, Martha G. Frank, Linda M. Hasselbach, and Dawn M. Littleton. 1989. Motivation and insight in wolf (*Canis lupus*) and Alaskan Malemute (*Canis familiaris*): visual discrimination and learning. *Bulletin of the Psychonomic Society* 27 (5):455–58.

Galton, Francis. 1908. Domestication of animals. In *Inquiries into Human Faculty and Development.* London: J. M. Dent and Co.

Goodmann, Patricia A., and Erich Klinghammer. 1985. Wolf ethogram. Battle Ground, Ind.: Wolf Park.

Gould, Stephen Jay, and Elisabeth Vrba. 1982. Exaptation—a missing term in the science of form. *Paleobiology* 8 (1):4–15.

Herre, W., and M. Rohrs. 1977. Origins of agriculture. In *World Anthropology,* edited by C. A. Reed. The Hague: Mouton.

Klinghammer, Erich, and Patricia Ann Goodmann. 1987. Socialization and management of wolves in captivity. In *Man and Wolf,* edited by H. Frank Dordrecht. The Netherlands: Dr. W. Junk Publishers.

Klinghammer, Erich. 1994. Imprinting and early experience: How to avoid problems with tame animals. Vol. 8, *Ethology Series.* Battle Ground, Ind.: North American Wildlife Park Foundation.

Kohane, M. J., and P .A. Parsons. 1988. Domestication: Evolutionary change under stress. *Evolutionary Biology* 23:31–48.

Lewin, Roger. 1988. A revolution of ideas in agricultural origins. *Science* 240:984–86.

Manwell, Clyde, and C. M. Ann Baker. 1983. Origin of the dog: From wolf or wild *Canis familiaris. Speculations in Science and Technology* 6 (3):213–24.

Meggitt, M. J. 1965. The association between Australian Aborigines and dingos. In *Man, Culture, and Animals,* edited by A. Leeds and A. P. Vayda. Washington, D.C.: American Association for the Advancement of Science.

Oppenheimer, E. C., and J. R. Oppenheimer. 1975. Certain behavioral features in the pariah dog (*Canis familiaris*) in West Bengal. *Applied Animal Ethology* 2:81–92.

Parsons, Peter A. 1997. Stress-resistance genotypes, metabolic efficiency and interpreting evolutionary change. In *Environmental Stress, Adaptation and Evolution,* edited by R. Bijlsma and V. Loeschcke. Boston: Birkhäuser.

Price, Edward O. 1998. Behavioral genetics and the process of animal domestication. In *Genetics and the Behavior of Domestic Animals,* edited by Temple Grandin. San Diego: Academic Press.

Scott, John Paul. 1968. Evolution and domestication of the dog. *Evolutionary Biology* 2:243–75.

Scott, John P. 1959. The inheritance of annual breeding cycles in hybrid basenji–cocker spaniel dogs. *The Journal of Heredity,* 255–61.

Smith, Charles K. 1970. Logical and persuasive structures in Charles Darwin's prose style. *Language and Style,* Fall:243–73.

Trut, Lyudmila N. 1999. Early canid domestication: the farm-fox experiment. *American Scientist,* March–April, 160–69.

Woolpy, Jerome H., and Benson E. Ginsburg. 1967. Wolf socialization: A study of temperament in a wild social species. *American Zoologist* 7:357–63.

Zeuner, Frederick E. 1963. *A History of Domesticated Animals.* New York and Evanston: Harper and Row.

Zimen, Erik. 1981. *The Wolf: A Species in Danger.* New York: Delacorte Press.

———. 1999. An animal changes the world: the domestication of the wolf. Paper read at Wolf & Co.:International Symposium on Canids, at Cologne, Germany.

TWO

Beck, Alan M. 1973. *The Ecology of Stray Dogs: a Study of Free-ranging Urban Animals.* Baltimore: York Press.

Bekoff, M. 1977. Mammalian dispersal and the ontogeny of individual behavioural phenotypes. *American Naturalist* 111:715–32.

Blumler, Mark, and Roger Byrne. 1991. The ecological genetics of domestication and the origins of agriculture. *Current Anthropology* 32 (1):23–54.

Boitani, Luigi, Francesco Francisci, Paolo Ciucci, and Giorgio Andreoli. 1995. Population biology and ecology of feral dogs in central Italy. In *The Domestic Dog: Its Evolution, Behaviour, and Interactions with People,* edited by J. Serpell. Cambridge, UK: Cambridge University Press.

Carr, G. M., and D. W. Macdonald. 1986. The sociality of solitary foragers: A model based on resource dispersion. *Animal Behaviour* 35:1540–49.

Corbett, Laurie. 1995. *The Dingo in Australia and Asia.* Ithaca: Comstock/Cornell.

Daniels, Thomas J. 1983. The social organization of free-ranging urban dogs. I. Nonestrous social behavior. *Applied Animal Ethology* 10:341–63.

Daniels, Thomas J., and Marc Bekoff. 1989a. Population and social biology of free-ranging dogs, *Canis familiaris. Journal of Mammalogy* 70 (4):754–62.

———. 1989b. Spatial and temporal resource use by feral and abandoned dogs. *Ethology* 81:300–312.

Feldmann, Bruce. 1974. The problem with urban dogs. *Science* 185 (4155):3.

Frank, H., and M. G. Frank. 1982. On the effects of domestication on canine social development and behavior. *Applied Animal Ethology* 8:507–25.

Lantis, Margaret. 1981. Zoonotic diseases in the Canadian and Alaskan North. *Inuit Studies* 5 (2):83–107.

Macdonald, David W., and Geoff M. Carr. 1995. Variation in dog society: Between resource dispersion and social flux. In *The Domestic Dog: Its Evolution, Behaviour, and Interactions with People,* edited by J. Serpell. Cambridge, UK: Cambridge University Press.

Plyusnina, I. Z., and L. N. Trut. 1991. An analysis of fear and aggression during early development of behavior in silver foxes (*Vulpes vulpes*). *Applied Animal Behaviour Science* 32:253–68.

Yeager, Rodger, and Norman N. Miller. 1986. *Wildlife, Wild Death.* Edited by L. W. Milbrath, SUNY Series in Environmental Public Policy. Albany, N.Y.:State University of New York.

THREE

Ash, Edward C. 1972 [1927]. *Dogs: Their History and Development.* Reissue ed. Two vols. New York: Benjamin Blom.

Brisbin, I. Lehr. 1977. The pariah: Its ecology and importance to the origin, development and study of pure bred dogs. *Pure Bred Dogs: American Kennel Club Gazette,* January, 27–29.

Clutton–Brock, Juliet. 1977. Man-made dogs. *Science* 197:1340–1342.

Coppinger, Raymond, and Richard Schneider. 1995. Evolution of working dogs. In *The Domestic Dog: Its Evolution, Behaviour, and Interactions with People,* edited by J. Serpell. Cambridge, UK: Cambridge University Press.

Coppinger, Raymond, and Lorna Coppinger. 1998. Differences in the behavior of dog breeds. In *Genetics and the Behavior of Domestic Animals,* edited by Temple Grandin. San Diego: Academic Press.

Das, Gautam. 1998. Notes on the Indian "Pariah" dog (Indian Spitz) as a hunting dog. *Merigal,* October, 12–13.

Meyrick, John. 1861. *House Dogs and Sporting Dogs: Their Varieties, Points, Management, Training, Breeding, Rearing and Diseases.* London: John Van Voorst.

Stonehenge. 1872. *The Dog in Health and Disease.* Second edition. London: Longmans, Green, Reader, and Dyer.

Titcomb, M. 1969. Dogs and man in the ancient Pacific with special attention to Hawaii. *Special Publication 59,* Bernice P. Bishop Museum, Honolulu, Hawaii.

FOUR

Andelt, William F. 1992. Effectiveness of livestock guarding dogs for reducing predation on domestic sheep. *Wildlife Society Bulletin* 20:55–62.

Bailey, Ed. 1997/98. The fearful dog. *Gun Dog,* December/January, 41–44.

———. 1998. Another look at the 49th day. *Gun Dog,* April/May, 20–24.

———. 1999. Giving pups a head start. *Gun Dog,* April/May, 24–27.

Bateson, Gregory. 1972. Metalogue: What is an instinct? In *Steps to an Ecology of Mind.* New York: Ballantine Books.

Bateson, Patrick. 1979. How do sensitive periods arise and what are they for? *Animal Behaviour* 27:470–86.

Black, Hal L. 1981. Navajo sheep and goat guarding dogs: A new world solution to the coyote problem. *Rangelands* 3 (6):235–38.

Black, H. L., and J. S. Green. 1985. Navajo use of mixed–breed dogs for management of predators. *Journal of Range Management* 38:11–15.

Cairns, Robert B. 1966. Development, maintenance and extinction of social attachment behavior in sheep. *Journal of Comparative and Physiological Psychology* 62 (2):298–306.

Cairns, Robert B., and Donald L. Johnson. 1965. The development of interspecies social attachments. *Psychonometric Science* 2:337–38.

Cairns, Robert B., and Jack Werboff. 1967. Behavior development in the dog: An interspecific analysis. *Science* 158:1070–72.

Cooper, R.M., and John P. Zubek. 1958. Effects of enriched and restricted early environments on the learning ability of bright and dull rats. *Canadian Journal of Psychology* 12 (3):159–64.

Coppinger, L., and R. Coppinger. 1993. Dogs for herding and guarding livestock. In *Livestock Handling and Transport,* edited by T. Grandin. Wallingford, Oxford, UK: CAB International.

Coppinger, Raymond, Lorna Coppinger, Gail Langeloh, Lori Gettler, and Jay Lorenz. 1988. A decade of use of livestock guarding dogs. *In Proceedings of the Vertebrate Pest Conference,* edited by A. C. Crabb and R. E. Marsh. Davis, CA: University of California.

Cornwell, Anne C., and John L. Fuller. 1961. Conditioned responses in young puppies. *Journal of Comparative and Physiological Psychology* 54 (1):13–15.

Cronly–Dillon, John. 1982. The experience that shapes our brains. *New Scientist,* November, 366–69.

Elliot, Orville, and J. P. Scott. 1961. The development of emotional distress reactions to separation in puppies. *The Journal of Genetic Psychology* 99:3–32.

Fairfax, Harrison, ed. 1913 *Roman Farm Management: The Treatises of Cato and Varro Done into English, with Notes of Modern Instances, by a Virginia Farmer.* New York: Macmillan.

Leyhausen, Paul. 1973. On the function of the relative hierarchy of moods. In *Motivation of Human and Animal Behavior: An Ethological View,* edited by K. Lorenz and P. Leyhausen. New York: Van Nostrand Reinhold.

Markowitz, Tim M., Martin R. Dally, Karin Gursky, and Edward Price. 1998. Early handling increases lamb affinity for humans. *Animal Behaviour* 55:573–87.

Pfaffenberger, Clarence J. 1963. *The New Knowledge of Dog Behavior.* New York: Howell Book House.

Scott, J. P., and Mary Vesta Marston. Critical periods affecting the development of normal and mal-adjustive behavior in puppies. *Journal of Genetics and Psychology* 77:25ff.

Wahlsten, Douglas. 1990. Insensitivity of the analysis of variance to heredity–environment interaction. *Behavioral and Brain Sciences* 13 (1):109–20.

FIVE

Adelman, Steven, Richard Taylor, and Norman Heglund. 1975. Sweating on paws and palms: What is its function? *American Journal of Physiology* 229 (5):1400–02.

Baker, M. A., and L .W. Chapman. 1977. Rapid brain cooling in exercising dogs. *Science* 195:781–83.

Coppinger, Lorna. 1977. *The World of Sled Dogs.* New York: Howell Book House.

Coppinger, Ray, and Carleton Phillips. 1981. Temperature regulations in running sled dogs. *Team and Trail,* May, 10–12.

Fox, M. W., and J. W. Spencer. 1969. Exploratory behavior in the dog: experiential or age dependent? *Developmental Psychobiology* 2 (2):68–74.

Lantis, Margaret. 1980. Changes in the Alaskan Eskimo relation of man to dog and their effect on two human diseases. *Arctic Anthropology* 17 (1):2–24.

Lawton, Terry Lee. 1975. A biomechanical analysis of an Alaskan husky sled dog racing team. Doctorate, Exercise Science, University of Massachusetts, Amherst.

MacRury, Ian Kenneth. 1991. The Inuit dog: Its provenance, environment and history. Master of Philosophy, Scott Polar Research Institute, University of Cambridge, Darwin College, Cambridge, UK.

Paulsen, Gary. 1994. *Winterdance: The Fine Madness of Running the Iditarod.* San Diego: Harcourt Brace.

Phillips, Carleton J., Raymond Coppinger, and David Schimel. 1981. Hyperthermia in running sled dogs. *Journal of Applied Phsysiology: Respiratory, Environmental and Exercise Physiology* 51 (1):135–42.

Sands, Michael W., Raymond P. Coppinger, and Carleton J. Phillips. 1977. Comparisons of thermal sweating and histology of sweat glands of selected canids. *Journal of Mammalogy* 58 (1):74–78.

Schmidt–Nielsen, Knut, William Bretz, and Richard Taylor. 1970. Panting in dogs: Unidirectional air flow over evaporative surfaces. *Science* 169:1102–04.

Taylor, C. R., Knut Schmidt–Nielsen, Razi Dmi'el, and Michael Fedak. 1971. Effect of hyperthermia on heat balance during running in the African hunting dog. *American Journal of Physiology* 220 (3):823–27.

SIX

Arons, Cynthia D., and William J. Shoemaker. 1992. The distribution of catecholamines and β-endorphin in the brains of three behaviorally distinct breeds of dogs and their F–1 hybrids. *Brain Research* 594:31–39.

Bekoff, Marc. 1974. Social play and play–soliciting by infant canids. *American Zoologist* 14:323–40.

———. 1989. Social play and physical training: when "not enough" may be plenty. *Ethology* 80:330–33.

Bekoff, Marc, Harriet L. Hill, and Jeffry Mitton. 1975. Behavioral taxonomy in canids by discriminant function analyses. *Science* 190:1223–24.

Bleicher, Norman. 1962. Behavior of the bitch during parturition. *Journal of the American Veterinary Medical Association* 140 (10):1076–82.

———. 1963. Physical and behavioral analysis of dog vocalizations. *American Journal of Veterinary Research* 224 (100):415–26.

Boissy, Alain. 1998. Fear and fearfulness in determining behavior. In *Genetics and the Behavior of Domestic Animals,* edited by T. Grandin. San Diego: Academic Press.

Coppinger, R. P., C. K. Smith, and L. Miller. 1985. Observations on why mongrels may make effective livestock protecting dogs. *Journal of Range Management* 38 (6):560–61.

Coren, Stanley. 1994. *The Intelligence of Dogs.* New York: Bantam Books.

Eisenberg, J. F., and Paul Leyhausen. 1972. The phylogenesis of predatory behavior in mammals. *Zeitschrift für Tierpsychologie* 30:59–93.

Fox, M. W. 1969. Ontogeny of prey-killing behavior in *Canidae. Behavior* 35 (18): 259–70.

Freedman, D. G. 1958. Constitutional and environmental interactions in rearing of four breeds of dogs. *Science* 127:585–86.

Hahn, Martin E., and John C. Wright. 1998. The influence of genes on social behavior of dogs. In *Genetics and the Behavior of Domestic Animals,* edited by T. Grandin. San Diego: Academic Press.

Hall, W. G., and Christina L. Williams. 1983. Suckling isn't feeding, or is it? A search for developmental continuities. *Advances in the Study of Behavior* 13:219–55.

Hirsch, Jerry. 1964. Breeding analysis of natural units in behavior genetics. *American Zoologist* 4:139–45.

Klinghammer, Erich. 1976. Wolf ethogram—abbreviations and definitions. Battle Ground, Ind.: Wolf Park.

Mahut, Helen. 1958. Breed differences in the dog's emotional behaviour. *Canadian Journal of Psychology* 12 (1):35–44.

McConnell, P. A., and J. R. Bayliss. 1985. Interspecific communication in cooperative herding: Acoustic and visual signals from human shepherds and herding dogs. *Zeitschrift für Tierpsychologie* 67:303–28.

Moore, George, and Theresa Moore. 1976. Herding instincts, trials and your stock dog. *The World of the Working Dog* 2 (4):14–19.

Vincent, Louis E., and Marc Bekoff. 1978. Quantitative analyses of the ontogeny of predatory behaviour in coyotes, *Canis latrans. Animal Behaviour* 26:225–31.

SEVEN

Bateson, Patrick. 1991. Assessment of pain in animals. *Animal Behaviour* 42:827–39.

Beck, Alan M. 1973. *The Ecology of Stray Dogs: A Study of Free-ranging Urban Animals.* Baltimore: York Press.

Beck, Alan, and Aaron Katcher. 1996. *Between Pets and People.* West Lafayette, Ind.: Purdue University Press.

Blaisdell, John D. 1999. The rise of man's best friend: The popularity of dogs as companion animals in late eighteenth century London as reflected by the dog tax of 1796. *Anthrozoös* 12 (2):76–87.

Borchelt, Peter L., and Victoria L. Voith. 1982a. Classification of animal behavior problems. *Veterinary Clinics of North America: Small Animal Practice* 12:571–85.

———. 1982b. Diagnosis and treatment of separation–related behavior problems in dogs. *Veterinary Clinics of North America: Small Animal Practice* 12:625–35.

Brown, C. J., O. D. Murphree, and E. O. Newton. 1978. The effect of inbreeding on human aversion in pointer dogs. *The Journal of Heredity* 69:362–65.

Coppinger, Raymond. 1996. *Fishing Dogs.* Berkeley, Calif.: Ten Speed Press.

Dawkins, Marian Stamp. 1988. Behavioural deprivation: A central problem in animal welfare. *Applied Animal Behaviour Science* 20:209–25.

Dodman, Nicholas. 1996. *The Dog Who Loved Too Much.* New York: Bantam.

Elliot, Orville, and J. P. Scott. 1961. The development of emotional distress reactions to separation in puppies. *The Journal of Genetic Psychology* (99):3–32.

Estep, Daniel Q. 1996. The ontogeny of behavior. In *Readings in Companion Animal Behavior,* edited by V. L. Voith and P. L. Borchelt. Trenton, N.J.: Veterinary Learning Systems.

Fisher, John. 1995. *Understanding the Behaviour of the Pet Dog.* Tisbury, Wiltshire, England: COAPE.

Freedman, D. G. 1958. Constitutional and environmental interactions in rearing of four breeds of dogs. *Science* 127:585–86.

Freedman, Daniel G., John A. King, and Orville Elliot. 1960. Critical period in the social development of dogs. *Science* 133:1016–17.

Hart, Benjamin L., and Lynette A. Hart. 1985. Selecting pet dogs on the basis of cluster analysis of breed behavior profiles and gender. *Journal of the American Veterinary Medical Association* 186 (11):1181–85.

Hart, Benjamin L., and Michael F. Miller. 1985. Behavioral profiles of dog breeds. *Journal of the American Veterinary Medical Association* 186 (11):1175–80.

Houpt, Katherine A. 1985. Companion animal behavior: A review of dog and cat behavior in the field, the laboratory and the clinic. *Cornell Veterinary* 75:248–61.

Jones, Mitchell. 1988. *The Dogs of Capitalism. Book 1: Origins.* Austin, Tex.: 21st Century Logic.

McCrave, Elizabeth. 1991. Diagnostic criteria for separation anxiety in the dog. *Veterinary Clinics of North America: Small Animal Practice* 21 (2):247–55.

Radde, Glenn. 1980. The significance of the dog: Perspectives of dog human relationships. Masters, Department of Geography, University of Minnesota.

Serpell, James. 1986. *In the Company of Animals.* Oxford, UK: Basil Blackwell.

Thomson, Keith. 1996. The rise and fall of the English bulldog. *American Scientist* 84:220–23.

Voith, Victoria, and Peter Borchelt. 1982. Diagnosis and treatment of dominance aggression in dogs. *Veterinary Clinics of North America: Small Animal Practice* 12 (4):655–63.

Voith, Victoria L., and Peter L. Borchelt, eds. 1996. *Readings in Companion Animal Behavior. Trenton*, N.J.: Veterinary Learning Systems.

EIGHT

Allen, Karen, and Jim Blascovich. 1996. The value of service dogs for people with severe ambulatory disabilities. *Journal of the American Medical Association* 275 (13):1001–06.

Baun, Mara M., Nancy Bergstrom, Nancy F. Langston, and Linda Thoma. 1984. Physiological effects of the human/companion animal bond. *Nursing Research* 33 (3):126–29.

Coppinger, Raymond, and Justine Lyons. 1995. *Wheelchair Assistance Dogs*. Amherst, Mass.: Hampshire College.

Coppinger, Raymond, Lorna Coppinger, and Ellen Skillings. 1998. Observations on assistance dog training and use. *Journal of Applied Animal Welfare Science* 1 (2):133–44.

Coppinger, R., and J. Zuccotti. 1999. Exercise and socialization in dogs. *Journal of Applied Animal Welfare Science* 2:281–96.

Eddy, Jane, Lynette A. Hart, and Ronald P. Boltz. 1988. The effects of service dogs on social acknowledgments of people in wheelchairs. *The Journal of Psychology* 122 (1):39–45.

Fuller, John L. 1960. Behavior genetics. *Annual Review of Psychology* 11:41–70.

———. 1967. Experiential deprivation and later behavior. *Science* 158:1645–52.

Gagnon, Sylvain, and Francois Y. Doré. 1992. Search behavior in various breeds of adult dogs (*Canis familiaris*): Object permanence and olfactory cues. *Journal of Comparative Psychology* 106 (1):58–68.

Goddard, M. E., and R. G. Beilharz. 1982/83. Genetic traits which determine the suitability of dogs as guide-dogs for the blind. *Applied Animal Ethology* 9:299–315.

———. 1986. Early prediction of adult behaviour in potential guide dogs. *Applied Animal Behaviour Science* 15:247–60.

Hart, Lynette A, Benjamin L. Hart, and Bonita Bergin. 1987. Socializing effects of service dogs for people with disabilities. *Anthrozoös* 1:41–44.

Hart, Lynette A., R. Lee Zasloff, and Anne Marie Benfoatto. 1995. The pleasures and problems of hearing dog ownership. *Psychological Reports* 77:969–70.

Hirsch, Jerry. 1990. A nemesis for heritability estimation. *Behavioral and Brain Sciences* 13 (1):137–38.

Leighton, Eldin. 1997. Genetics of canine hip dysplasia. *Journal of the American Veterinary Medical Association* 210 (10):1474–79.

Mader, Bonnie, Lynette A. Hart, and Bonita Bergin. 1989. Social acknowledgments for children with disabilities: Effects of service dogs. *Child Development* 60:1529–34.

Mitchell, Daniel S. 1976. *Selection of Dogs for Land Mine and Booby Trap Detection Training*. Fort Belvoir, Virginia: U.S. Army Mobility Equipment Research and Development Command.

O'Farrell, V. 1986. *Manual of Canine Behaviour*. Cheltenham, Gloucestershire, UK: British Small Animal Veterinary Association.

Pfaffenberger, Clarence J. 1963. *The New Knowledge of Dog Behavior*. New York: Howell Book House.

Pfaffenberger, Clarence J., and John Paul Scott. 1976. Early rearing and testing. In *Guide Dogs for the Blind: Their Selection, Development and Training*, edited by C. Pfaffenberger, J. P. Scott, J. Fuller, B. Ginsburg, and S. Bielfelt. New York: Elsevier Scientific Publishing Company.

Serpell, James, Raymond Coppinger, and Aubrey H. Fine. 2000. The welfare of assistance and therapy animals: An ethical comment. In *Handbook on Animal Assisted Therapy*, edited by A. H. Fine. San Diego: Academic Press.

Struckus, Joseph Edward. 1989. The use of pet-facilitated therapy in the treatment of depression in the elderly: A behavioral conceptualization of treatment effect. Doctorate, Psychology, University of Massachusetts, Amherst.

NINE

Brisbin, Jr., I. Lehr, Raymond P. Coppinger, Mark H. Feinstein, Steven N. Austad, and John J. Mayer. 1994. The New Guinea singing dog: Taxonomy, captive studies and conservation priorities. *Science in New Guinea* 20 (1):27–38.

Darwin, Charles. 1859. Letter to Asa Gray. *Journal of the Proceedings of the Linnean Society (Zoology)* 3:50–53.

Lamarck, J. B. 1984 [1809]. *Zoological Philosophy: An Exposition with Regard to the Natural History of Animals*. Chicago: University of Chicago Press.

Mayr, Ernst. 1942. *Systematics and the Origin of Species*. New York: Columbia University Press.

———. 1996. The modern evolutionary theory. *Journal of Mammalogy* 77 (1):1–7.

Honacki, J. H., K. E. Kinman, and J. W. Koeppl, eds. 1982. *Mammal Species of the World: A Taxonomic and Geographic Reference*. Lawrence, Kansas: Allen Press and Association of Systematic Collections.

Lorenz, Konrad. 1954. *Man Meets Dog*. Boston: Houghton Mifflin Company.

Troughton, E. 1957. A new native dog from the Papuan Highlands. *Proceedings of the Royal Zoological Society of New South Wales* 1955–1956:93–94.

TEN

Barnosky, Anthony D. 1990. Punctuated equilibrium and phyletic gradualism. In *Current Mammalogy*, edited by H. Genoways. New York: Plenum Press.

Clark, Kate M. 1996. Neolithic dogs: A reappraisal based on evidence from the remains of a large canid deposited in a ritual feature. *International Journal of Osteoarchaeology* 6:211–19.

Clutton-Brock, Juliet. 1992. The process of domestication. *Mammal Review* 22 (2):79–85.

———. 1995. Origins of the dog: Domestication and early history. In *The Domestic Dog, Its Evolution, Behavior, and Interactions with People*, edited by J. Serpell. Cambridge, UK: Cambridge University Press.

Ferrell, Robert, Donald Morizot, Jacqueline Horn, and Curtis Carley. 1980. Biochemical markers in a species endangered by introgression: the Red Wolf. *Biochemical Genetics* 18 (1/2):39–49.

Geist, Valerius. 1986. On speciation in Ice Age mammals, with special reference to cervids and caprids. *Canadian Journal of Zoology* 65:1067–84.

Kurtén, Björn. 1968. *Pleistocene Mammals of Europe*. London: Weidenfeld and Nicolson.

Kurtén, Björn, and E. Anderson. 1980. *Pleistocene Mammals of North America*. New York: Columbia University Press.

Lehman, Niles, Andrew Eisenhawer, Kimberly Hansen, L. David Mech, Rolf O. Peterson, Peter J. P. Gogan, and Robert K. Wayne. 1991. Introgression of coyote mitochondrial DNA into sympatric North American grey wolf populations. *Evolution* 45 (1):104–19.

Randi, E., V. Lucchini, M. F. Christensen, N. Mucci, S. M. Funk, G. Dolf, and V. Loeschcke. 2000. Mitochondrial DNA variability in Italian and East European wolves. *Conservation Biology* 14: 464–73.

Vilà, Carles, and Robert Wayne. 1999. Hybridization between wolves and dogs. *Conservation Biology* 13 (1):195–98.

Vilà, O. C., P. Savolainen, J. E. Malconado, I. R. Amorim, J. E. Rice, R. L. Honeycutt, K. A. Crandall, J. Lundeberg, and R. K.Wayne. 1997. Multiple and ancient origins of the domestic dog. *Science* 276:1687–89.

Wayne, R. K., N. Lehman, M. W. Allard, and R. L. Honeycutt. 1992. Mitochondrial DNA variability of the gray wolf: Genetic consequences of population decline and habitat fragmentation. *Conservation Biology* 6:559–69.

ELEVEN

Alberch, P. 1982a. Developmental constraints in evolutionary processes. In *Evolution and Development,* edited by J. T. Bonner. New York: Springer-Verlag.

Alberch, Pere. 1982b. The generative and regulatory roles of development in evolution. Edited by D. Mossakowsky and G. Roth, *Environmental Adaptation and Evolution.* Stuttgart and New York: Gustav Fischer.

Alberch, Pere, Stephen J. Gould, George F. Oster, and David B. Wake. 1979. Size and shape in ontogeny and phylogeny. *Paleobiology* 5:296–317.

Clutton-Brock, J., G. B. Corbet, and M. Hills. 1976. A review of the family *Canidae* with a classification by numerical methods. *Bulletin of the British Museum (Natural History) (Zoology)* 29:117–99.

Coppinger, Raymond, and Lorna Coppinger. 1996. Biologic bases of behavior of domestic dog breeds. In *Readings in Companion Animal Behavior,* edited by V. Voith and P. Borchelt. Trenton, N.J.: Veterinary Learning Systems.

Godfrey, Laurie R., and Michael R. Sutherland. 1995. What's growth got to do with it? Process and product in the evolution of ontogeny. *Journal of Human Evolution* 29:405–31.

Gould, Stephen Jay. 1966. Allometry and size in ontogeny and phylogeny. *Biological Review* 41:587–640.

———. 1986. The egg-a-day barrier. *Natural History* 86(7):16–24.

Haldane, J. B. S. 1930. *The Causes of Evolution.* London: Longmans Green.

Hildebrand, Milton. 1952. An analysis of body proportions in the *Canidae. American Journal of Anatomy* 90:217–56.

Kruska, D. 1988. Mammalian domestication and its effects on brain structure and behaviour. In *Intelligence and Evolutionary Biology,* edited by H. J. Jerison and I. Jerison. Berlin: Springer-Verlag.

Lumer, Hyman. 1940. Evolutionary allometry in the skeleton of the domestic dog. *The American Naturalist* 74:439–67.

Martin, Robert D. 1996. Scaling of the mammalian brain: The maternal energy hypothesis. *News in Physiological Sciences* 11:149–56.

Morey, Darcy F. 1992. Size, shape and development in the evolution of the domestic dog. *Journal of Archeological Science* 19:181–284.

Northcutt, R. G. 1990. Ontogeny and phylogeny: A re-evaluation of conceptual relationships and some applications. *Brain Behaviour and Evolution* 36:116–40.

Purves, Dale. 1988. Effects of animal size and form on the organization of the nervous system. In *Body and Brain: A Trophic Theory of Neural Connections.* Cambridge, Mass.: Harvard University Press.

Shea, Brian T. 1985. Ontogenetic allometry and scaling. In *Size and Scaling in Primate Biology,* edited by W. L. Jungers. New York: Plenum.

Stebbins, G. Ledyard, Jr. 1950. *Variation and Evolution in Plants.* New York: Columbia University Press.

———. 1959. The role of hybridization in evolution. *Proceedings of the American Philosophical Society* 103:231–51.

Stockard, Charles R. 1941. *The Genetic and Endocrinic Basis for Differences in Form and Behavior.* Philadelphia: The Wistar Institute of Anatomy and Biology.

Twitty, V.C. 1966. *Of Scientists and Salamanders.* San Francisco, Calif.: W. H. Freeman.

Wayne, Robert K. 1986. Cranial morphology of domestic and wild canids: The influence of development on morphological change. *Evolution* 40 (2):243–61.

Willis, Malcolm B. 1991. *The German Shepherd Dog: A Genetic History.* London: H. F. and G. Witherby Ltd.

Index